Plant Anatomy

Plant Anatomy

Edited by
David Bonham

Larsen & Keller
www.larsen-keller.com

Plant Anatomy
Edited by David Bonham
ISBN: 978-1-63549-646-8 (Hardback)

Larsen & Keller

Published by Larsen and Keller Education,
5 Penn Plaza,
19th Floor,
New York, NY 10001, USA

Cataloging-in-Publication Data

Plant anatomy / edited by David Bonham.
    p. cm.
Includes bibliographical references and index.
ISBN 978-1-63549-646-8
1. Plant anatomy. 2. Anatomy. 3. Botany. I. Bonham, David.
QK641 .P53 2018
581.4--dc23

For more information regarding Larsen and Keller Education and its products, please visit the publisher's website www.larsen-keller.com

Table of Contents

Preface

Plant anatomy is a vast subject which deals with the study of the cell structure, and tissues of plants at a microscopic level. It focuses on seed anatomy, flower anatomy, root structure, seed structure, leaf anatomy, wood anatomy, etc. While understanding the long-term perspectives of the topics, the book makes an effort in highlighting their impact as a modern tool for the growth of the discipline. The topics covered in it offer the readers new insights in the field of plant anatomy. For someone with an interest and eye for detail, this textbook covers the most significant topics in this field.

A foreword of all chapters of the book is provided below:

Chapter 1 - Plants are eukaryotic organisms that belong to the kingdom Plantae. Plant anatomy studies the structure of plants. The subject can be divided into various categories like stem anatomy, wood anatomy, root anatomy and seed anatomy. This chapter has been carefully written to provide an easy understanding of the varied facets of plant anatomy; **Chapter 2** - Plants are multicellular and plant cells contain a membrane within a nucleus. They have several distinctive features found in their membrane which cannot be found in other organisms. The two cell types are parenchyma cells and collenchyma cells. Plant cells are best understood in confluence with the major topics listed in the following chapter; **Chapter 3** - Plastids are double membrane organelle found in plants and selected algae. Chloroplast, chromoplast, leucoplast and gerontoplast are some categories of plastids. They have the ability to change into other cells depending on the function they play. The major categories of plastids are dealt with great details in the chapter; **Chapter 4** - Plants can be structurally divided into flowers, leaves, bark and seeds. A flower can further be divided into petals, sepal, gynoecium and stamen whereas a bark constitutes of phloem, cork cambium, cortex, vascular cambium and root. This chapter discusses the methods of plant anatomy in a critical manner providing key analysis to the subject matter; **Chapter 5** - Plant secretory tissues can be divided into lactiferous tissues and glandular tissues. Along with plant secretory tissues, the other tissues discussed are vascular tissues, spongy tissues and ground tissues. The aspects elucidated in this chapter are of vital importance, and provide a better understanding of plant anatomy.

At the end, I would like to thank all the people associated with this book devoting their precious time and providing their valuable contributions to this book. I would also like to express my gratitude to my fellow colleagues who encouraged me throughout the process.

Editor

An Introduction to Plant Anatomy

Plants are eukaryotic organisms that belong to the kingdom Plantae. Plant anatomy studies the structure of plants. The subject can be divided into various categories like stem anatomy, wood anatomy, root anatomy and seed anatomy. This chapter has been carefully written to provide an easy understanding of the varied facets of plant anatomy.

Plant Anatomy

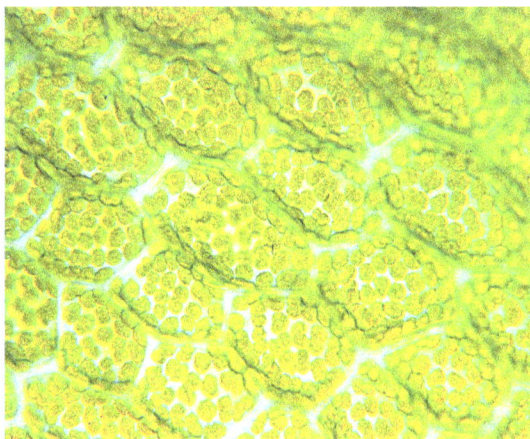

Chloroplasts in leaf cells of the moss *Mnium stellare*

Plant anatomy or phytotomy is the general term for the study of the internal structure of plants. While originally it included plant morphology, which is the description of the physical form and external structure of plants, since the mid-20th century the investigations of plant anatomy are considered a separate, distinct field, and plant anatomy refers to *just* the internal plant structures. Plant anatomy is now frequently investigated at the cellular level, and often involves the sectioning of tissues and microscopy.

Structural Divisions

Plant anatomy is often divided into the following categories:

Flower anatomy

Calyx

Corolla

Androecium

Gynoecium

Leaf anatomy

Stem anatomy

    Stem structure

Fruit/Seed anatomy

    Ovule

        Seed structure

    Pericarp

    Accessory fruit

Wood anatomy

    Bark

        Cork

        Phloem

        Vascular cambium

    Heartwood and sapwood

        branch collarw

Root anatomy

    Root structure

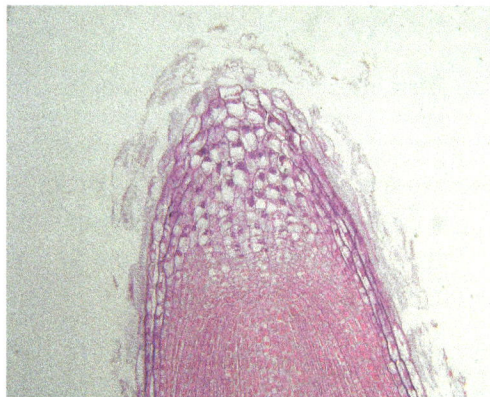

Root tip

History

About 300 BC Theophrastus wrote a number of plant treatises, only two of which survive, *Enquiry into Plants* and *On the Causes of Plants*. He developed concepts of plant morphology and classification, which did not withstand the scientific scrutiny of the Renaissance.

A Swiss physician and botanist, Gaspard Bauhin, introduced binomial nomenclature into plant taxonomy. He published *Pinax theatri botanici* in 1596, which was the first to use this convention for naming of species. His criteria for classification included natural relationships, or 'affinities', which in many cases were structural.

It was in the late 1600s that plant anatomy became refined into a modern science. Italian doctor and microscopist, Marcello Malpighi, was one of the two founders of plant anatomy. In 1671 he published his *Anatomia Plantarum*, the first major advance in plant physiogamy since Aristotle. The other founder was the British doctor Nehemiah Grew. He published *An Idea of a Philosophical History of Plants* in 1672 and *The Anatomy of Plants* in 1682. Grew is credited with the recognition of plant cells, although he called them 'vesicles' and 'bladders'. He correctly identified and described the sexual organs of plants (flowers) and their parts.

In the Eighteenth Century, Carl Linnaeus established taxonomy based on structure, and his early work was with plant anatomy. While the exact structural level which is to be considered to be scientifically valid for comparison and differentiation has changed with the growth of knowledge, the basic principles were established by Linnaeus. He published his master work, *Species Plantarum* in 1753.

In 1802, French botanist Charles-François Brisseau de Mirbel, published *Traité d'anatomie et de physiologie végétale* (*Treatise on Plant Anatomy and Physiology*) establishing the beginnings of the science of plant cytology.

In 1812, Johann Jacob Paul Moldenhawer published *Beyträge zur Anatomie der Pflanzen*, describing microscopic studies of plant tissues.

In 1813 a Swiss botanist, Augustin Pyrame de Candolle, published *Théorie élémentaire de la botanique*, in which he argued that plant anatomy, not physiology, ought to be the sole basis for plant classification. Using a scientific basis, he established structural criteria for defining and separating plant genera.

In 1830, Franz Meyen published *Phytotomie*, the first comprehensive review of plant anatomy.

In 1838 German botanist Matthias Jakob Schleiden, published *Contributions to Phytogenesis*, stating, "the lower plants all consist of one cell, while the higher plants are composed of (many) individual cells" thus confirming and continuing Mirbel's work.

A German-Polish botanist, Eduard Strasburger, described the mitotic process in plant cells and further demonstrated that new cell nuclei can only arise from the division of other pre-existing nuclei. His *Studien über Protoplasma* was published in 1876.

Gottlieb Haberlandt, a German botanist, studied plant physiology and classified plant tissue based upon function. On this basis, in 1884 he published *Physiologische Pflanzenanatomie* (*Physiological Plant Anatomy*) in which he described twelve types of tissue systems (absorptive, mechanical, photosynthetic, etc.).

British paleobotanists Dunkinfield Henry Scott and William Crawford Williamson described the structures of fossilized plants at the end of the Nineteenth Century. Scott's *Studies in Fossil Botany* was published in 1900.

Following Charles Darwin's *Origin of Species* a Canadian botanist, Edward Charles Jeffrey, who was studying the comparative anatomy and phylogeny of different vascular plant groups, applied the theory to plants using the form and structure of plants to establish a number of evolutionary lines. He published his *The Anatomy of Woody Plants* in 1917.

The growth of comparative plant anatomy was spearheaded by British botanist Agnes Arber. She published *Water Plants: A Study of Aquatic Angiosperms* in 1920, *Monocotyledons: A Morphological Study* in 1925, and *The Gramineae: A Study of Cereal, Bamboo and Grass* in 1934.

Following World War II, Katherine Esau published, *Plant Anatomy* (1953), which became the definitive textbook on plant structure in North American universities and elsewhere, it was still in print as of 2006. She followed up with her *Anatomy of seed plants* in 1960.

Plant

Plants are mainly multicellular, predominantly photosynthetic eukaryotes of the kingdom Plantae.

The term is today generally limited to the green plants, which form an unranked clade Viridiplantae (Latin for "green plants"). This includes the flowering plants, conifers and other gymnosperms, ferns, clubmosses, hornworts, liverworts, mosses and the green algae, and excludes the red and brown algae. Historically, plants formed one of two kingdoms covering all living things that were not animals, and both algae and fungi were treated as plants; however all current definitions of "plant" exclude the fungi and some algae, as well as the prokaryotes (the archaea and bacteria).

Green plants have cell walls containing cellulose and obtain most of their energy from sunlight via photosynthesis by primary chloroplasts, derived from endosymbiosis with cyanobacteria. Their chloroplasts contain chlorophylls a and b, which gives them their green color. Some plants are parasitic and have lost the ability to produce normal amounts of chlorophyll or to photosynthesize. Plants are characterized by sexual reproduction and alternation of generations, although asexual reproduction is also common.

There are about 300–315 thousand species of plants, of which the great majority, some 260–290 thousand, are seed plants. Green plants provide most of the world's molecular oxygen and are the basis of most of Earth's ecologies, especially on land. Plants that produce grains, fruits and vegetables form humankind's basic foodstuffs, and have been domesticated for millennia. Plants play many roles in culture. They are used as ornaments and, until recently and in great variety, they have served as the source of most medicines and drugs. The scientific study of plants is known as botany, a branch of biology.

Definition

Plants are one of the two groups into which all living things were traditionally divided; the other is animals. The division goes back at least as far as Aristotle (384 BC – 322 BC), who distinguished between plants, which generally do not move, and animals, which often are mobile to catch their food. Much later, when Linnaeus (1707–1778) created the basis of the modern system of scientific classification, these two groups became the kingdoms Vegetabilia (later Metaphyta or Plantae) and Animalia (also called Metazoa). Since then, it has become clear that the plant kingdom as originally defined included several unrelated groups, and the fungi and several groups of algae were removed to new kingdoms. However, these organisms are still often considered plants, particularly in popular contexts.

Outside of formal scientific contexts, the term "plant" implies an association with certain traits, such as being multicellular, possessing cellulose, and having the ability to carry out photosynthesis.

Current Definitions of Plantae

When the name Plantae or plant is applied to a specific group of organisms or taxon, it usually refers to one of four concepts. From least to most inclusive, these four groupings are:

Name(s)	Scope	Description
Land plants, also known as Embryophyta	Plantae *sensu strictissimo*	Plants in the strictest sense include the liverworts, hornworts, mosses, and vascular plants, as well as fossil plants similar to these surviving groups (e.g., Metaphyta Whittaker, 1969, Plantae Margulis, 1971).
Green plants, also known as Viridiplantae, Viridiphyta or Chlorobionta	Plantae *sensu stricto*	Plants in a strict sense include the green algae, and land plants that emerged within them, including stoneworts. The names given to these groups vary considerably as of July 2011. Viridiplantae encompass a group of organisms that have cellulose in their cell walls, possess chlorophylls *a* and *b* and have plastids that are bound by only two membranes that are capable of storing starch. It is this clade that is mainly the subject (e.g., Plantae Copeland, 1956).
Archaeplastida, also known as Plastida or Primoplantae	Plantae *sensu lato*	Plants in a broad sense comprise the green plants plus Rhodophyta (red algae) and Glaucophyta (glaucophyte algae). This clade includes the organisms that eons ago acquired their chloroplasts directly by engulfing cyanobacteria (e.g., Plantae Cavalier-Smith, 1981).
Old definitions of plant (obsolete)	Plantae *sensu amplo*	Plants in an ample sense refers to older, obsolete classifications that placed diverse algae, fungi or bacteria in Plantae (e.g., Plantae or Vegetabilia Linnaeus, Plantae Haeckel 1866, Metaphyta Haeckel, 1894, Plantae Whittaker, 1969).

Another way of looking at the relationships between the different groups that have been called "plants" is through a cladogram, which shows their evolutionary relationships. The evolutionary history of plants is not yet completely settled, but one accepted relationship between the three groups described above is shown below. Those which have been called "plants" are in bold.

The way in which the groups of green algae are combined and named varies considerably between authors.

Algae

Algae comprise several different groups of organisms which produce energy through photosynthesis and for that reason have been included in the plant kingdom in the past. Most conspicuous among the algae are the seaweeds, multicellular algae that may roughly resemble land plants, but are classified among the brown, red and green algae. Each of these algal groups also includes various microscopic and single-celled organisms. There is good evidence that some of these algal groups arose independently from separate non-photosynthetic ancestors, with the result that the brown algae, for example, are no longer classified within the plant kingdom as it is defined here.

Green algae from Ernst Haeckel's Kunstformen der Natur, 1904.

The Viridiplantae, the green plants – green algae and land plants – form a clade, a group consisting of all the descendants of a common ancestor. With a few exceptions among green algae, the green plants have the following features in common; cell walls containing cellulose, chloroplasts containing chlorophylls *a* and *b*, and food stores in the form of starch contained within the plastids. They undergo closed mitosis without centrioles, and typically have mitochondria with flat cristae. The chloroplasts of green plants are surrounded by two membranes, suggesting they originated directly from endosymbiotic cyanobacteria.

Two additional groups, the Rhodophyta (red algae) and Glaucophyta (glaucophyte algae), also have chloroplasts that appear to be derived directly from endosymbiotic cyanobacteria, although they differ in the pigments which are used in photosynthesis from those of the Viridiplantae and so are different in colour. In these groups, the storage polysaccharide is floridean starch and is stored in the cytoplasm rather than in the plastids. These groups appear to have had a common origin with Viridiplantae and the three groups form the clade Archaeplastida, whose name implies that their chloroplasts were derived from a single ancient endosymbiotic event. This is the broadest modern definition of the term 'plant'.

In contrast, most other algae (e.g. brown algae/diatoms, haptophytes, dinoflagellates, and euglenids) not only have different pigments but also have chloroplasts with three or four surrounding membranes. They are not close relatives of the Archaeplastida, presumably having acquired chloroplasts separately from ingested or symbiotic green and red algae. They are thus not included in even the broadest modern definition of the plant kingdom, although they were in the past.

The green plants or Viridiplantae were traditionally divided into the green algae (including the stoneworts) and the land plants. However, it is now known that the land plants evolved from within a group of green algae, so that the green algae by themselves are a paraphyletic group, i.e. a group that excludes some of the descendants of a common ancestor. Paraphyletic groups are generally avoided in modern classifications, so that in recent treatments the Viridiplantae have been divided into two clades, the Chlorophyta and the Streptophyta (including the land plants and Charophyta).

The Chlorophyta (a name that has also been used for *all* green algae) are the sister group to the group from which the land plants evolved. There are about 4,300 species of mainly marine organisms, both unicellular and multicellular. The latter include the sea lettuce, *Ulva*.

The other group within the Viridiplantae are the mainly freshwater or terrestrial Streptophyta, which consists of the land plants together with the Charophyta, itself consisting of several groups of green algae such as the desmids and stoneworts. Streptophyte algae are either unicellular or form multicellular filaments, branched or unbranched. The genus *Spirogyra* is a filamentous streptophyte alga familiar to many, as it is often used in teaching and is one of the organisms responsible for the algal "scum" that pond-owners so dislike. The freshwater stoneworts strongly resemble land plants and are believed to be their closest relatives. Growing in fresh water, they consist of a central stalk with whorls of branchlets, giving them a superficial resemblance to horsetails, species of the genus *Equisetum*, which are true land plants.

Fungi

The classification of fungi has been controversial until quite recently in the history of biology. Linnaeus' original classification placed the fungi within the Plantae, since they were unquestionably not animals or minerals and these were the only other alternatives. With later developments in microbiology, in the 19th century Ernst Haeckel felt that another kingdom was required to classify newly discovered micro-organisms. The introduction of the new kingdom Protista in addition to Plantae and Animalia, led to uncertainty as to whether fungi truly were best placed in the Plantae or whether they ought to be reclassified as protists. Haeckel himself found it difficult to decide and it was not until 1969 that a solution was found whereby Robert Whittaker proposed the creation of the kingdom Fungi. Molecular evidence has since shown that the most recent common ancestor (concestor), of the Fungi was probably more similar to that of the Animalia than to that of Plantae or any other kingdom.

Whittaker's original reclassification was based on the fundamental difference in nutrition between the Fungi and the Plantae. Unlike plants, which generally gain carbon through photosynthesis, and so are called autotrophs, fungi generally obtain carbon by breaking down and absorbing surrounding materials, and so are called heterotrophic saprotrophs. In addition, the substructure of multicellular fungi is different from that of plants, taking the form of many chitinous microscopic strands called hyphae, which may be further subdivided into cells or may form a syncytium containing many eukaryotic nuclei. Fruiting bodies, of which mushrooms are the most familiar example, are the reproductive structures of fungi, and are unlike any structures produced by plants.

Diversity

The table shows some species count estimates of different green plant (Viridiplantae) divisions. It suggests there are about 300,000 species of living Viridiplantae, of which 85–90% are flowering plants. (These are from different sources and different dates, they are not necessarily comparable, and like all species counts, are subject to a degree of uncertainty in some cases.)

Diversity of living green plant (Viridiplantae) divisions				
Informal group	Division name	Common name	No. of living species	Approximate No. in informal group
Green algae	Chlorophyta	green algae (chlorophytes)	3,800–4,300	8,500
	Charophyta	green algae (e.g. desmids & stoneworts)	2,800–6,000	(6,600–10,300)
Bryophytes	Marchantiophyta	liverworts	6,000–8,000	19,000
	Anthocerotophyta	hornworts	100–200	(18,100–20,200)
	Bryophyta	mosses	12,000	
Pteridophytes	Lycopodiophyta	club mosses	1,200	12,000
	Pteridophyta	ferns, whisk ferns & horse-tails	11,000	(12,200)
Seed plants	Cycadophyta	cycads	160	260,000
	Ginkgophyta	ginkgo	1	(259,511)
	Pinophyta	conifers	630	
	Gnetophyta	gnetophytes	70	
	Magnoliophyta	flowering plants	258,650	

The naming of plants is governed by the International Code of Nomenclature for algae, fungi, and plants and International Code of Nomenclature for Cultivated Plants.

Evolution

The evolution of plants has resulted in increasing levels of complexity, from the earliest algal mats, through bryophytes, lycopods, ferns to the complex gymnosperms and angiosperms of to-day. Plants in all of these groups continue to thrive, especially in the environments in which they evolved.

An algal scum formed on the land 1,200 million years ago, but it was not until the Ordovician Period, around 450 million years ago, that land plants appeared. However, new evidence from the study of carbon isotope ratios in Precambrian rocks has suggested that complex photosynthetic plants developed on the earth over 1000 m.y.a. For more than a century it has been assumed that the ancestors of land plants evolved in aquatic environments and then adapted to a life on land, an idea usually credited to botanist Frederick Orpen Bower in his 1908 book "The Origin of a Land Flora". A more recent alternative view, supported by genetic evidence, is that they evolved from single-celled algae that were already terrestrial. Primitive land plants began to diversify in the late Silurian Period, around 420 million years ago, and the fruits of their diversification are displayed in remarkable detail in an early Devonian fossil assemblage from the Rhynie chert. This chert preserved early plants in cellular detail, petrified in volcanic springs. By the middle of the Devonian Period most of the features recognised in plants today are present, including roots, leaves and secondary wood, and by late Devonian times seeds had evolved. Late Devonian plants had thereby reached a degree of sophistication that allowed them to form forests of tall trees. Evolutionary innovation continued after the Devonian period. Most plant groups were relatively unscathed by the Permo-Triassic extinction event, although the structures of communities changed. This may have set the scene for the evolution of flowering plants in the Triassic (~200 million years ago), which exploded in the Cretaceous and Tertiary. The latest major group of plants to evolve were the grasses, which became important in the mid Tertiary, from around 40 million years ago. The grasses, as well as many other groups, evolved new mechanisms of metabolism to survive the low CO_2 and warm, dry conditions of the tropics over the last 10 million years.

A 1997 proposed phylogenetic tree of Plantae, after Kenrick and Crane, is as follows, with modification to the Pteridophyta from Smith *et al*. The Prasinophyceae are a paraphyletic assemblage of early diverging green algal lineages, but are treated as a group outside the Chlorophyta: later authors have not followed this suggestion

Embryophytes

Dicksonia antarctica, a species of tree fern

The plants that are likely most familiar to us are the multicellular land plants, called embryophytes. Embryophytes include the vascular plants, such as ferns, conifers and flowering plants. They also include the *bryophytes*, of which mosses and liverworts are the most common.

All of these plants have eukaryotic cells with cell walls composed of cellulose, and most obtain their energy through photosynthesis, using light, water and carbon dioxide to synthesize food. About three hundred plant species do not photosynthesize but are parasites on other species of photosynthetic plants. Embryophytes are distinguished from green algae, which represent a mode of photosynthetic life similar to the kind modern plants are believed to have evolved from, by having specialized reproductive organs protected by non-reproductive tissues.

Bryophytes first appeared during the early Paleozoic. They can only survive where moisture is available for significant periods, although some species are desiccation-tolerant. Most species of bryophytes remain small throughout their life-cycle. This involves an alternation between two generations: a haploid stage, called the gametophyte, and a diploid stage, called the sporophyte. In bryophytes, the sporophyte is always unbranched and remains nutritionally dependent on its parent gametophyte. The bryophytes have the ability to secrete a cuticle on their outer surface, a waxy layer that confers resistant to desiccation. In the mosses and hornworts a cuticle is usually only produced on the sporophyte. Stomata are absent from liverworts, but occur on the sporangia of mosses and hornworts, allowing gas exchange while controlling water loss.

Vascular plants first appeared during the Silurian period, and by the Devonian had diversified and spread into many different terrestrial environments. They developed a number of adaptations that allowed them to spread into increasingly more arid places, notably the vascular tissues xylem and phloem, that transport water and food throughout the organism. Root systems capable of obtaining soil water and nutrients also evolved during the Devonian. In modern vascular plants, the sporophyte is typically large, branched, nutritionally independent and long-lived, but there is increasing evidence that Paleozoic gametophytes were just as complex as the sporophytes. The gametophytes of all vascular plant groups evolved to become reduced in size and prominence in the life cycle.

The first seed plants, Pteridosperms, now extinct, appeared in the Devonian and diversified through the Carboniferous. In these the microgametophyte is reduced to pollen and the megagametophyte remains inside the megasporangium, attached to the parent plant. A megasporangium invested in protective layer called an integument is known as an ovule. After fertilisation by means of sperm deposited by pollen grains, an embryo develops inside the ovule. The integument becomes a seed coat, and the ovule develops into a seed. Seed plants can survive and reproduce in extremely arid conditions, because they are not dependent on free water for the movement of sperm, or the development of free living gametophytes.

Early seed plants are gymnosperms, as the ovules and subsequent seeds are not enclosed in a protective structure (carpels or fruit), but are found naked, typically on cone scales. Pollen typically lands directly on the ovule. Four surviving groups remain widespread now, particularly the conifers, which are dominant trees in several biomes.

Fossils

Plant fossils include roots, wood, leaves, seeds, fruit, pollen, spores, phytoliths, and amber (the fossilized resin produced by some plants). Fossil land plants are recorded in terrestrial, lacustrine, fluvial and nearshore marine sediments. Pollen, spores and algae (dinoflagellates and acritarchs) are used for dating sedimentary rock sequences. The remains of fossil plants are not as common as fossil animals, although plant fossils are locally abundant in many regions worldwide.

The earliest fossils clearly assignable to Kingdom Plantae are fossil green algae from the Cambrian. These fossils resemble calcified multicellular members of the Dasycladales. Earlier Precambrian fossils are known that resemble single-cell green algae, but definitive identity with that group of algae is uncertain.

The oldest known fossils of embryophytes date from the Ordovician, though such fossils are fragmentary. By the Silurian, fossils of whole plants are preserved, including the lycophyte *Baragwanathia longifolia*. From the Devonian, detailed fossils of rhyniophytes have been found. Early fossils of these ancient plants show the individual cells within the plant tissue. The Devonian period also saw the evolution of what many believe to be the first modern tree, *Archaeopteris*. This fern-like tree combined a woody trunk with the fronds of a fern, but produced no seeds.

A petrified log in Petrified Forest National Park, Arizona

The Coal measures are a major source of Paleozoic plant fossils, with many groups of plants in existence at this time. The spoil heaps of coal mines are the best places to collect; coal itself is the remains of fossilised plants, though structural detail of the plant fossils is rarely visible in coal. In the Fossil Grove at Victoria Park in Glasgow, Scotland, the stumps of *Lepidodendron* trees are found in their original growth positions.

The fossilized remains of conifer and angiosperm roots, stems and branches may be locally abundant in lake and inshore sedimentary rocks from the Mesozoic and Cenozoic eras. Sequoia and its allies, magnolia, oak, and palms are often found.

Petrified wood is common in some parts of the world, and is most frequently found in arid or desert areas where it is more readily exposed by erosion. Petrified wood is often heavily silicified (the organic material replaced by silicon dioxide), and the impregnated tissue is often preserved in fine detail. Such specimens may be cut and polished using lapidary equipment. Fossil forests of petrified wood have been found in all continents.

Fossils of seed ferns such as *Glossopteris* are widely distributed throughout several continents of the Southern Hemisphere, a fact that gave support to Alfred Wegener's early ideas regarding Continental drift theory.

The earliest fossils attributed to green algae date from the Precambrian (ca. 1200 mya). The resistant outer walls of prasinophyte cysts (known as phycomata) are well preserved in fossil deposits of the Paleozoic (ca. 250-540 mya). A filamentous fossil (Proterocladus) from middle Neoproterozoic deposits (ca. 750 mya) has been attributed to the Cladophorales, while the oldest reliable records of the Bryopsidales, Dasycladales) and stoneworts are from the Paleozoic.

Structure, Growth and Development

Most of the solid material in a plant is taken from the atmosphere. Through a process known as photosynthesis, most plants use the energy in sunlight to convert carbon dioxide from the atmosphere, plus water, into simple sugars. (Parasitic plants, on the other hand, use the resources of its host to grow.) These sugars are then used as building blocks and form the main structural component of the plant. Chlorophyll, a green-colored, magnesium-containing pigment is essential to this process; it is generally present in plant leaves, and often in other plant parts as well.

The leaf is usually the primary site of photosynthesis in plants.

Plants usually rely on soil primarily for support and water (in quantitative terms), but also obtain compounds of nitrogen, phosphorus, potassium, magnesium and other elemental nutrients. Epiphytic and lithophytic plants depend on air and nearby debris for nutrients, and carnivorous plants supplement their nutrient requirements with insect prey that they capture. For the majority of plants to grow successfully they also require oxygen in the atmosphere and around their roots (soil gas) for respiration. Plants use oxygen and glucose (which may be produced from stored starch) to provide energy. Some plants grow as submerged aquatics, using oxygen dissolved in the surrounding water, and a few specialized vascular plants, such as mangroves, can grow with their roots in anoxic conditions.

Factors Affecting Growth

The genotype of a plant affects its growth. For example, selected varieties of wheat grow rapidly, maturing within 110 days, whereas others, in the same environmental conditions, grow more slowly and mature within 155 days.

Growth is also determined by environmental factors, such as temperature, available water, available light, carbon dioxide and available nutrients in the soil. Any change in the availability of these external conditions will be reflected in the plant's growth.

Biotic factors are also capable of affecting plant growth. Plants compete with other plants for space, water, light and nutrients. Plants can be so crowded that no single individual produces normal growth, causing etiolation and chlorosis. Optimal plant growth can be hampered by grazing animals, suboptimal soil composition, lack of mycorrhizal fungi, and attacks by insects or plant diseases, including those caused by bacteria, fungi, viruses, and nematodes.

Simple plants like algae may have short life spans as individuals, but their populations are commonly seasonal. Other plants may be organized according to their seasonal growth pattern: annual plants live and reproduce within one growing season, biennial plants live for two growing seasons and usually reproduce in second year, and perennial plants live for many growing seasons and continue to reproduce once they are mature. These designations often depend on climate and other environmental factors; plants that are annual in alpine or temperate regions can be biennial or perennial in warmer climates. Among the vascular plants, perennials include both evergreens that keep their leaves the entire year, and deciduous plants that lose their leaves for some part of it. In temperate and boreal climates, they generally lose their leaves during the winter; many tropical plants lose their leaves during the dry season.

There is no photosynthesis in deciduous leaves in autumn.

The growth rate of plants is extremely variable. Some mosses grow less than 0.001 millimeters per hour (mm/h), while most trees grow 0.025-0.250 mm/h. Some climbing species, such as kudzu, which do not need to produce thick supportive tissue, may grow up to 12.5 mm/h.

Plants protect themselves from frost and dehydration stress with antifreeze proteins, heat-shock proteins and sugars (sucrose is common). LEA (Late Embryogenesis Abundant) protein expression is induced by stresses and protects other proteins from aggregation as a result of desiccation and freezing.

Effects of Freezing

When water freezes in plants, the consequences for the plant depend very much on whether the freezing occurs within cells (intracellularly) or outside cells in intercellular spaces (Glerum 1985). Intracellular freezing, which usually kills the cell (Lyons et al. 1979) regardless of the hardiness of the plant and its tissues, seldom occurs in nature because rates of cooling are rarely high enough to support it. Rates of cooling of several degrees Celsius per minute are typically needed to cause intracellular formation of ice (Mazur 1977).

At rates of cooling of a few degrees Celsius per hour, segregation of ice occurs in intercellular spaces, the "extraorgan ice" of Sakai and Larcher (1987) and their coworkers. This may or may not be lethal, depending on the hardiness of the tissue.

The process of intercellular ice formation was described by Glerum (1985). At freezing temperatures, water in the intercellular spaces of plant tissue freezes first, though the water may remain unfrozen until temperatures drop below −7 °C (19 °F). After the initial formation of ice intercellularly, the cells shrink as water is lost to the segregated ice, and the cells undergo freeze-drying. This dehydration is now considered the fundamental cause of freezing injury.

DNA Damage and Repair

Plants are continuously exposed to a range of biotic and abiotic stresses. These stresses often cause

DNA damage directly, or indirectly via the generation of reactive oxygen species. Plants are capable of a DNA damage response that is a critical mechanism for maintaining genome stability. The DNA damage response is particularly important during seed germination, since seed quality tends to deteriorate with age in association with DNA damage accumulation. During germination repair processes are activated to deal with this accumulated DNA damage. In particular, single- and double-strand breaks in DNA can be repaired. The DNA checkpoint kinase ATM has a key role in integrating progression through germination with repair responses to the DNA damages accumulated by the aged seed.

Plant Cells

Plant cells are typically distinguished by their large water-filled central vacuole, chloroplasts, and rigid cell walls that are made up of cellulose, hemicellulose, and pectin. Cell division is also characterized by the development of a phragmoplast for the construction of a cell plate in the late stages of cytokinesis. Just as in animals, plant cells differentiate and develop into multiple cell types. Totipotent meristematic cells can differentiate into vascular, storage, protective (e.g. epidermal layer), or reproductive tissues, with more primitive plants lacking some tissue types.

Physiology

Photosynthesis

Plants are photosynthetic, which means that they manufacture their own food molecules using energy obtained from light. The primary mechanism plants have for capturing light energy is the pigment chlorophyll. All green plants contain two forms of chlorophyll, chlorophyll a and chlorophyll b. The latter of these pigments is not found in red or brown algae. The simple equation of photosynthesis is as follows:-

$6CO_2 + 6H_2O \rightarrow$ (in the presence of light and chlorophyll) $C_6H_{12}O_6 + 6O_2$

Immune System

By means of cells that behave like nerves, plants receive and distribute within their systems information about incident light intensity and quality. Incident light that stimulates a chemical reaction in one leaf, will cause a chain reaction of signals to the entire plant via a type of cell termed a *bundle sheath cell*. Researchers, from the Warsaw University of Life Sciences in Poland, found that plants have a specific memory for varying light conditions, which prepares their immune systems against seasonal pathogens. Plants use pattern-recognition receptors to recognize conserved microbial signatures. This recognition triggers an immune response. The first plant receptors of conserved microbial signatures were identified in rice (XA21, 1995) and in *Arabidopsis thaliana* (FLS2, 2000). Plants also carry immune receptors that recognize highly variable pathogen effectors. These include the NBS-LRR class of proteins.

Internal Distribution

Vascular plants differ from other plants in that nutrients are transported between their different parts through specialized structures, called xylem and phloem. They also have roots for taking up water and minerals. The xylem moves water and minerals from the root to the rest of the plant, and the phloem provides the roots with sugars and other nutrient produced by the leaves.

Ecology

The photosynthesis conducted by land plants and algae is the ultimate source of energy and organic material in nearly all ecosystems. Photosynthesis radically changed the composition of the early Earth's atmosphere, which as a result is now 21% oxygen. Animals and most other organisms are aerobic, relying on oxygen; those that do not are confined to relatively rare anaerobic environments. Plants are the primary producers in most terrestrial ecosystems and form the basis of the food web in those ecosystems. Many animals rely on plants for shelter as well as oxygen and food.

Land plants are key components of the water cycle and several other biogeochemical cycles. Some plants have coevolved with nitrogen fixing bacteria, making plants an important part of the nitrogen cycle. Plant roots play an essential role in soil development and prevention of soil erosion.

Distribution

Plants are distributed worldwide in varying numbers. While they inhabit a multitude of biomes and ecoregions, few can be found beyond the tundras at the northernmost regions of continental shelves. At the southern extremes, plants have adapted tenaciously to the prevailing conditions.

Plants are often the dominant physical and structural component of habitats where they occur. Many of the Earth's biomes are named for the type of vegetation because plants are the dominant organisms in those biomes, such as grasslands and forests.

Ecological Relationships

Numerous animals have coevolved with plants. Many animals pollinate flowers in exchange for food in the form of pollen or nectar. Many animals disperse seeds, often by eating fruit and passing the seeds in their feces. Myrmecophytes are plants that have coevolved with ants. The plant provides a home, and sometimes food, for the ants. In exchange, the ants defend the plant from herbivores and sometimes competing plants. Ant wastes provide organic fertilizer.

The majority of plant species have various kinds of fungi associated with their root systems in a kind of mutualistic symbiosis known as mycorrhiza. The fungi help the plants gain water and mineral nutrients from the soil, while the plant gives the fungi carbohydrates manufactured in photosynthesis. Some plants serve as homes for endophytic fungi that protect the plant from herbivores by producing toxins. The fungal endophyte, *Neotyphodium coenophialum*, in tall fescue (*Festuca arundinacea*) does tremendous economic damage to the cattle industry in the U.S.

Various forms of parasitism are also fairly common among plants, from the semi-parasitic mistletoe that merely takes some nutrients from its host, but still has photosynthetic leaves, to the fully parasitic broomrape and toothwort that acquire all their nutrients through connections to the roots of other plants, and so have no chlorophyll. Some plants, known as myco-heterotrophs, parasitize mycorrhizal fungi, and hence act as epiparasites on other plants.

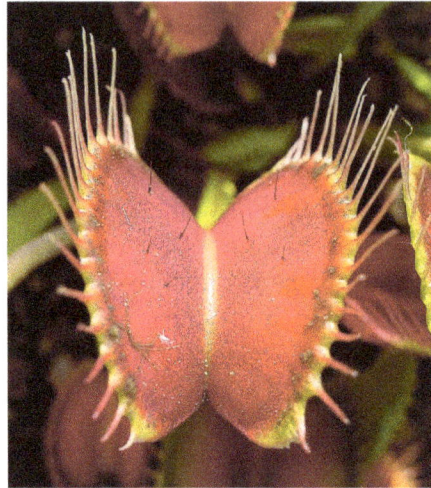

The Venus flytrap, a species of carnivorous plant.

Many plants are epiphytes, meaning they grow on other plants, usually trees, without parasitizing them. Epiphytes may indirectly harm their host plant by intercepting mineral nutrients and light that the host would otherwise receive. The weight of large numbers of epiphytes may break tree limbs. Hemiepiphytes like the strangler fig begin as epiphytes but eventually set their own roots and overpower and kill their host. Many orchids, bromeliads, ferns and mosses often grow as epiphytes. Bromeliad epiphytes accumulate water in leaf axils to form phytotelmata that may contain complex aquatic food webs.

Approximately 630 plants are carnivorous, such as the Venus Flytrap (*Dionaea muscipula*) and sundew (*Drosera* species). They trap small animals and digest them to obtain mineral nutrients, especially nitrogen and phosphorus.

Importance

The study of plant uses by people is termed economic botany or ethnobotany; some consider economic botany to focus on modern cultivated plants, while ethnobotany focuses on indigenous plants cultivated and used by native peoples. Human cultivation of plants is part of agriculture, which is the basis of human civilization. Plant agriculture is subdivided into agronomy, horticulture and forestry.

Foods and Beverages

Much of human nutrition depends on plants, either directly through foods and beverages consumed by people, or indirectly as feed for animals or the flavoring of foods. The science of agriculture deals with the planting, raising, nutrition, and harvest of food crops, and has played a key role in the history of world civilizations.

Human nutrition depends to a large extent on cereals, especially maize (or corn), wheat, rice, oats, and millet. Large areas of many countries are given over to the cultivation of cereals for local consumption or export to other countries. Livestock animals including cows, pigs, sheep, goats and camels are all herbivores; and most feed primarily or entirely on cereal plants. Cereals are staple crops, meaning that they provide calories (in the form of complex carbohydrates such as starch) that are needed to fuel daily activities, and thus form the foundation of a daily diet. Other staple crops include potatoes, cassava, yams, and legumes.

Mechanical harvest of oats.

Human food also includes vegetables, which consist principally of leaves and stems eaten as food. Vegetables are important for the vitamins, minerals, and dietary fiber they supply. Fruits provide a higher quantity of sugars and have a sweeter taste than vegetables. However, whether a particular food is considered a "vegetable" or a "fruit" will depend on context, since the word *fruit* has a more precise definition in botany than it does in general use. Nuts and seeds, including foods such as peanuts, walnuts, almonds, and pistachios, contain unsaturated fats that are also necessary for a healthy diet. As with fruits, the terms *nut* and *seed* have stricter definitions in plant science.

Many plants are used to flavor foods. Such plants include herbs (e.g. rosemary and mint), which come from the green leafy parts of plants, and spices (e.g. cumin and cinnamon), which come from other plant parts. Some plants produce edible flowers, which may be added to salads or used to decorate foods. Sweeteners such as sugar and stevia are derived from plants. Sugar is obtained mainly from sugar cane and sugar beet, and honey is created when bees regurgitate the nectar from flowers. Cooking oils and margarine come from maize, soybean, rapeseed, safflower, sunflower, olive and others. Food additives include gum arabic, guar gum, locust bean gum, starch and pectin.

Plants are also the source of beverages produced either by infusion, such as coffee and tea; by fermentation, such as beer and wine; or by distillation, such as whisky, vodka, rum, and other alcoholic spirits.

Nonfood Products

Plants are the source of many natural products such as essential oils, natural dyes, pigments, waxes, resins, tannins, alkaloids, amber and cork. Products derived from plants include soaps, shampoos, perfumes, cosmetics, paint, varnish, turpentine, rubber, latex, lubricants, linoleum, plastics, inks, and gums. Renewable fuels from plants include firewood, peat and many other biofuels. Coal and petroleum are fossil fuels derived from the remains of plants. Olive oil has been used in lamps for centuries to provide illumination.

Timber in storage for later processing at a sawmill.

Structural resources and fibers from plants are used in both the construction of dwellings and the manufacture of clothing. Wood is used not only for buildings, boats, and furniture, but also for smaller items such as musical instruments and sports equipment. Wood also may be pulped for the manufacture of paper and cardboard. Cloth is often made from cotton, flax, ramie or synthetic fibers derived from cellulose, such as rayon and acetate. The thread that is used to sew cloth likewise comes from plant fibers. Hemp and jute are grown for their fibers, which may be woven into rope or rough sacking.

Plants are also a primary source of basic chemicals, both for their medicinal and physiological effects, as well as for the industrial synthesis of a vast array of organic chemicals. Medicines derived from plants include aspirin, taxol, morphine, quinine, reserpine, colchicine, digitalis and vincristine. There are hundreds of herbal supplements such as ginkgo, Echinacea, feverfew, and Saint John's wort. Pesticides derived from plants include nicotine, rotenone, strychnine and pyrethrins. Certain plants contain psychotropic chemicals that are extracted and ingested, including tobacco, cannabis (marijuana), opium, and cocaine. Poisons from plants include ricin, hemlock and curare.

Aesthetic Uses

Thousands of plant species are cultivated for aesthetic purposes as well as to provide shade, modify temperatures, reduce wind, abate noise, provide privacy, and prevent soil erosion. Plants are the basis of a multibillion-dollar per year tourism industry, which includes travel to historic gardens, national parks, rainforests, forests with colorful autumn leaves, and the National Cherry Blossom Festival.

A rose espalier at Niedernhall in Germany.

Capitals of ancient Egyptian columns decorated to resemble papyrus plants (at Luxor, Egypt).

While some gardens are planted with food crops, many are planted for aesthetic, ornamental, or conservation purposes. Arboretums and botanical gardens are public collections of living plants. In private outdoor gardens, lawn grasses, shade trees, ornamental trees, shrubs, vines, herbaceous perennials and bedding plants are used. Gardens may cultivate the plants in a naturalistic state, or may sculpture their growth, as with topiary or espalier. Gardening is the most popular leisure activity in the U.S., and working with plants or horticulture therapy is beneficial for rehabilitating people with disabilities.

Plants may also be grown or kept indoors as houseplants, or in specialized buildings such as greenhouses that are designed for the care and cultivation of living plants. Venus Flytrap, sensitive plant and resurrection plant are examples of plants sold as novelties. There are also art forms specializing in the arrangement of cut or living plant, such as bonsai, ikebana, and the arrangement of cut or dried flowers. Ornamental plants have sometimes changed the course of history, as in tulipomania.

Architectural designs resembling plants appear in the capitals of ancient Egyptian columns, which were carved to resemble either the Egyptian white lotus or the papyrus. Images of plants are often used in painting and photography, as well as on textiles, money, stamps, flags and coats of arms.

Scientific and Cultural Uses

Barbara McClintock (1902–1992) was a pioneering cytogeneticist
who used maize (or corn) to study the mechanism of inheritance of traits.

Basic biological research has often been done with plants. In genetics, the breeding of pea plants allowed Gregor Mendel to derive the basic laws governing inheritance, and examination of chromosomes in maize allowed Barbara McClintock to demonstrate their connection to inherited traits. The plant *Arabidopsis thaliana* is used in laboratories as a model organism to understand how genes control the growth and development of plant structures. Space stations or space colonies may one day rely on plants for life support.

Ancient trees are revered and many are famous. Tree rings themselves are an important method of dating in archeology, and serve as a record of past climates.

Plants figure prominently in mythology, religion and literature. They are used as national and state emblems, including state trees and state flowers. Plants are often used as memorials, gifts and to mark special occasions such as births, deaths, weddings and holidays. The arrangement of flowers may be used to send hidden messages.

The field of ethnobotany studies plant use by indigenous cultures, which helps to conserve endangered species as well as discover new medicinal plants.

Negative Effects

Weeds are uncultivated and usually unwanted plants growing in managed environments such as farms, urban areas, gardens, lawns, and parks. People have spread plants beyond their native ranges and some of these introduced plants become invasive, damaging existing ecosystems by displacing native species. Invasive plants cause costly damage in crop losses annually by displacing crop plants, they further increase the cost of production and the use of chemicals to control them, which in turn affects the environment.

Plants may cause harm to animals, including people. Plants that produce windblown pollen invoke allergic reactions in people who suffer from hay fever. A wide variety of plants are poisonous. Toxalbumins are plant poisons fatal to most mammals and act as a serious deterrent to consumption. Several plants cause skin irritations when touched, such as poison ivy. Certain plants contain psychotropic chemicals, which are extracted and ingested or smoked, including tobacco, cannabis (marijuana), cocaine and opium. Smoking causes damage to health or even death, while some drugs may also be harmful or fatal to people. Both illegal and legal drugs derived from plants may have negative effects on the economy, affecting worker productivity and law enforcement costs. Some plants cause allergic reactions when ingested, while other plants cause food intolerances that negatively affect health.

References

- Kenrick, Paul; Crane, Peter R. (1997). The origin and early diversification of land plants: A cladistic study. Washington, D. C.: Smithsonian Institution Press. ISBN 1-56098-730-8

- Lewis, L.A.; McCourt, R.M. (2004). "Green algae and the origin of land plants". American Journal of Botany. 91: 1535–1556. PMID 21652308. doi:10.3732/ajb.91.10.1535

- Butterfield, Nicholas J.; Knoll, Andrew H.; Swett, Keene (1994). "Paleobiology of the Neoproterozoic Svanbergfjellet Formation, Spitsbergen". Lethaia. 27 (1): 76–76. ISSN 0024-1164. doi:10.1111/j.1502-3931.1994.tb01558.x

- Raven, Peter H.; Evert, Ray F.; Eichhorn, Susan E. (2005). Biology of Plants (7th ed.). New York: W. H. Freeman and Company. ISBN 0-7167-1007-2

- Adl, S.M. et al. (2005). "The new higher level classification of eukaryotes with emphasis on the taxonomy of protists". Journal of Eukaryote Microbiology. 52: 399–451. PMID 16248873. doi:10.1111/j.1550-7408.2005.00053.x

- Waterworth WM, Bray CM, West CE (2015). "The importance of safeguarding genome integrity in germination and seed longevity". J. Exp. Bot. 66 (12): 3549–58. PMID 25750428. doi:10.1093/jxb/erv080

- Van den Hoek, C., D. G. Mann, & H. M. Jahns, 1995. Algae: An Introduction to Phycology. pages 343, 350, 392, 413, 425, 439, & 448 (Cambridge: Cambridge University Press). ISBN 0-521-30419-9

- Goyal, K., Walton, L. J., & Tunnacliffe, A. (2005). "LEA proteins prevent protein aggregation due to water stress". Biochemical Journal. 388 (Part 1): 151–157. PMC 1186703. PMID 15631617. doi:10.1042/BJ20041931. Archived from the original on 3 August 2009

- Song, W.Y.; et al. (1995). "A receptor kinase-like protein encoded by the rice disease resistance gene, XA21". Science. 270 (5243): 1804–1806. PMID 8525370. doi:10.1126/science.270.5243.1804

- Gifford, Ernest M.; Foster, Adriance S. (1988). Morphology and Evolution of Vascular Plants (3rd ed.). New York: W. H. Freeman and Company. p. 358. ISBN 0-7167-1946-0

- Guiry, M.D. & Guiry, G.M. (2007). "Phylum: Chlorophyta taxonomy browser". AlgaeBase version 4.2 Worldwide electronic publication, National University of Ireland, Galway. Retrieved 2007-09-23

- Novíkov & Barabaš-Krasni (2015). "Modern plant systematics". Liga-Pres: 685. ISBN 978-966-397-276-3. doi:10.13140/RG.2.1.4745.6164

- Whittaker, R. H. (1969). "New concepts of kingdoms or organisms" (PDF). Science. 163 (3863): 150–160. PMID 5762760. doi:10.1126/science.163.3863.150

An Overview of Plant Cell

Plants are multicellular and plant cells contain a membrane within a nucleus. They have several distinctive features found in their membrane which cannot be found in other organisms. The two cell types are parenchyma cells and collenchyma cells. Plant cells are best understood in confluence with the major topics listed in the following chapter.

Plant Cell

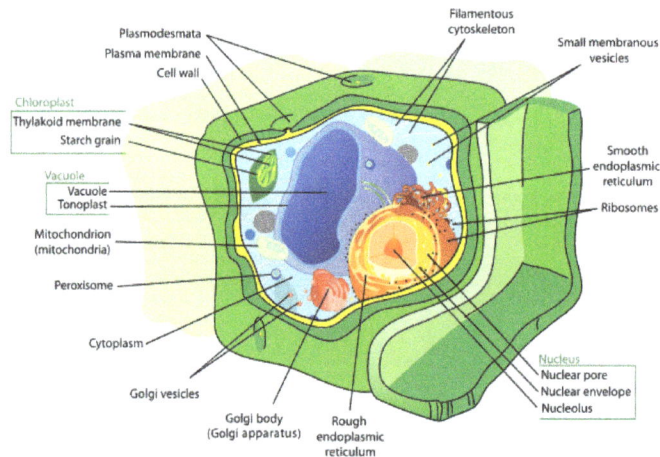

Plant cell structure

Plant cells are eukaryotic cells that differ in several key aspects from the cells of other eukaryotic organisms. Their distinctive features include:

- A large central vacuole, a water-filled volume enclosed by a membrane known as the *tonoplast* that maintains the cell's turgor, controls movement of molecules between the cytosol and sap, stores useful material and digests waste proteins and organelles.

- A cell wall composed of cellulose and hemicelluloses, pectin and in many cases lignin, is secreted by the protoplast on the outside of the cell membrane. This contrasts with the cell walls of fungi, which are made of chitin, and of bacteria, which are made of peptidoglycan. Cell walls perform many essential functions: they provide shape to form the tissue and organs of the plant, and play an important role in intercellular communication and plant-microbe interactions

- Specialized cell-to-cell communication pathways known as plasmodesmata, pores in the primary cell wall through which the plasmalemma and endoplasmic reticulum of adjacent cells are continuous.

- Plastids, the most notable being the chloroplast, which contains chlorophyll, a green-colored pigment that absorbs sunlight, and allows the plant to make its own food in the process known as photosynthesis. Other types of plastids are the amyloplasts, specialized for starch storage, elaioplasts specialized for fat storage, and chromoplasts specialized for synthesis and storage of pigments. As in mitochondria, which have a genome encoding 37 genes, plastids have their own genomes of about 100–120 unique genes and, it is presumed, arose as prokaryotic endosymbionts living in the cells of an early eukaryotic ancestor of the land plants and algae.

- Cell division by construction of a phragmoplast as a template for building a cell plate late in cytokinesis is characteristic of land plants and a few groups of algae, notably the Charophytes and the Chlorophyte Order Trentepohliales

- The motile, free-swimming sperm of bryophytes and pteridophytes, cycads and *Ginkgo* are the only cells of land plants to have flagella similar to those in animal cells, but the conifers and flowering plants do not have motile sperm and lack both flagella and centrioles.

Cell Types

- Parenchyma cells are living cells that have functions ranging from storage and support to photosynthesis and phloem loading (transfer cells). Apart from the xylem and phloem in their vascular bundles, leaves are composed mainly of parenchyma cells. Some parenchyma cells, as in the epidermis, are specialized for light penetration and focusing or regulation of gas exchange, but others are among the least specialized cells in plant tissue, and may remain totipotent, capable of dividing to produce new populations of undifferentiated cells, throughout their lives. Parenchyma cells have thin, permeable primary walls enabling the transport of small molecules between them, and their cytoplasm is responsible for a wide range of biochemical functions such as nectar secretion, or the manufacture of secondary products that discourage herbivory. Parenchyma cells that contain many chloroplasts and are concerned primarily with photosynthesis are called chlorenchyma cells. Others, such as the majority of the parenchyma cells in potato tubers and the seed cotyledons of legumes, have a storage function.

- Collenchyma cells – collenchyma cells are alive at maturity and have only a primary wall. These cells mature from meristem derivatives that initially resemble parenchyma, but differences quickly become apparent. Plastids do not develop, and the secretory apparatus (ER and Golgi) proliferates to secrete additional primary wall. The wall is most commonly thickest at the corners, where three or more cells come in contact, and thinnest where only two cells come in contact, though other arrangements of the wall thickening are possible.

Pectin and hemicellulose are the dominant constituents of collenchyma cell walls of dicotyledon angiosperms, which may contain as little as 20% of cellulose in *Petasites*. Collenchyma cells are typically quite elongated, and may divide transversely to give a septate appearance. The role of this cell type is to support the plant in axes still growing in length, and to confer flexibility and tensile strength on tissues. The primary wall lacks lignin that would make it tough and rigid, so this cell type provides what could be called plastic support – support that can hold a young stem or petiole into the air, but in cells that can be stretched as the cells around them elongate. Stretchable support (without elastic snap-back) is a good way to describe what collenchyma does. Parts of the strings in celery are collenchyma.

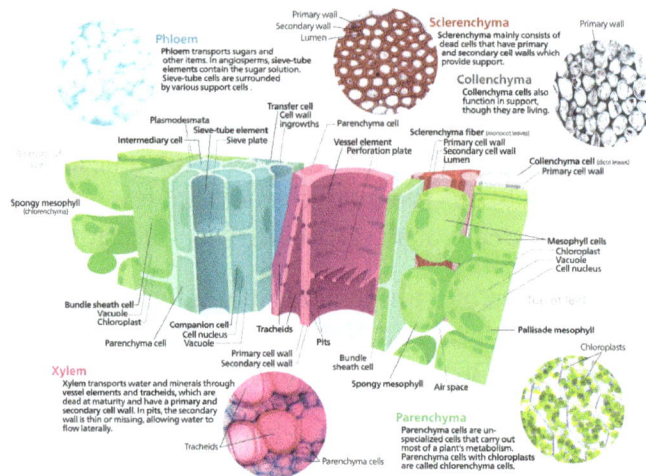

Cross section of a leaf showing various plant cell types

- Sclerenchyma cells – Sclerenchyma cells (from the Greek skleros, *hard*) are hard and tough cells with a function in mechanical support. They are of two broad types – sclereids or stone cells and fibres. The cells develop an extensive secondary cell wall that is laid down on the inside of the primary cell wall. The secondary wall is impregnated with lignin, making it hard and impermeable to water. Thus, these cells cannot survive for long' as they cannot exchange sufficient material to maintain active metabolism. Sclerenchyma cells are typically dead at functional maturity, and the cytoplasm is missing, leaving an empty central cavity.

Functions for sclereid cells (hard cells that give leaves or fruits a gritty texture) include discouraging herbivory, by damaging digestive passages in small insect larval stages, and physical protection (a solid tissue of hard sclereid cells form the pit wall in a peach and many other fruits). Functions of fibres include provision of load-bearing support and tensile strength to the leaves and stems of herbaceous plants. Sclerenchyma fibres are not involved in conduction, either of water and nutrients (as in the xylem) or of carbon compounds (as in the phloem), but it is likely that they may have evolved as modifications of xylem and phloem initials in early land plants.

Tissue Types

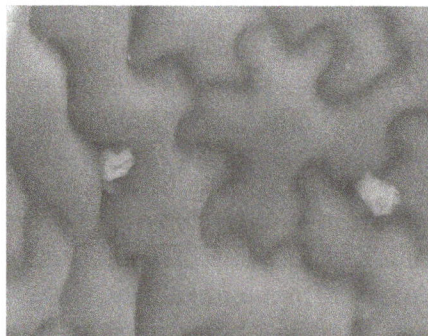

Cells of *Arabidopsis thaliana* epidermis

The major classes of cells differentiate from undifferentiated meristematic cells (analogous to the stem cells of animals) to form the tissue structures of roots, stems, leaves, flowers, and reproductive structures.

Xylem cells are elongated cells with lignified secondary thickening of the cell walls. Xylem cells are specialised for conduction of water, and first appeared in plants during their transition to land in the Silurian period more than 425 million years ago. The possession of xylem defines the vascular plants or Tracheophytes. Xylem tracheids are pointed, elongated xylem cells, the simplest of which have continuous primary cell walls and lignified secondary wall thickenings in the form of rings, hoops, or reticulate networks. More complex tracheids with valve-like perforations called bordered pits characterise the gymnosperms. The ferns and other pteridophytes and the gymnosperms have only xylem tracheids, while the angiosperms also have xylem vessels. Vessel members are hollow xylem cells without end walls that are aligned end-to-end so as to form long continuous tubes. The bryophytes lack true xylem cells, but their sporophytes have a water-conducting tissue known as the hydrome that is composed of elongated cells of simpler construction.

Phloem is a specialised tissue for food transport in higher plants. Phloem cells mainly transport sucrose along pressure gradients generated by osmosis. This phenomenon is called translocation. Phloem consists of two cell types, the sieve tubes and the intimately associated companion cells. The sieve tube elements lack nuclei and ribosomes, and their metabolism and functions are regulated by the adjacent nucleate companion cells. Sieve tubes are joined end-to-end with perforate end-plates between known as *sieve plates*, which allow transport of photosynthate between the sieve elements. The companion cells, connected to the sieve tubes via plasmodesmata, are responsible for loading the phloem with sugars. The bryophytes lack phloem, but moss sporophytes have a simpler tissue with analogous function known as the leptome.

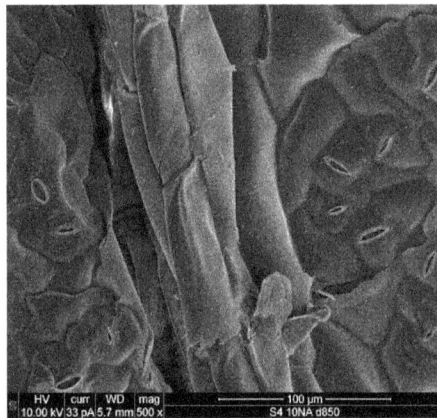

This is an electron micrograph of the epidermal cells of a
Brassica chinensis leaf. The stomates are also visible.

Plant epidermal cells are specialised parenchyma cells covering the external surfaces of leaves, stems and roots. The epidermal cells of aerial organs arise from the superficial layer of cells known as the *tunica* (L1 and L2 layers) that covers the plant shoot apex, whereas the cortex and vascular tissues arise from innermost layer of the shoot apex known as the *corpus* (L3 layer). The epidermis of roots originates from the layer of cells immediately beneath the root cap.

The epidermis of all aerial organs, but not roots, is covered with a cuticle made of the polyester cutin and/or the hydrocarbon polymer cutan with a superficial layer of epicuticular waxes. The epidermal cells of the primary shoot are thought to be the only plant cells with the biochemical capacity to synthesize cutin. Several cell types may be present in the epidermis. Notable among these are the stomatal guard cells, glandular and clothing hairs or trichomes, and the root hairs of primary roots. In the shoot epidermis of most plants, only the guard cells have chloroplasts. Chloroplasts contain the green pigment chlorophyll which is needed for photosynthesis.

Organelles

- Cell membrane
- Cell wall
- Nuclear membrane
- Vacuole
- Plastid
- Chloroplast
- Leucoplast

- Chromoplast
- Golgi Bodies
- Cytoplasm
- Nucleus
- Chromatin
- Cytoskeleton
- Nucleolus

Cell Wall

A cell wall is a structural layer surrounding some types of cells, situated outside the cell membrane. It can be tough, flexible, and sometimes rigid. It provides the cell with both structural support and protection, and also acts as a filtering mechanism. Cell walls are present in most prokaryotes (except mycoplasma bacteria), in algae, plants and fungi but rarely in other eukaryotes including animals. A major function is to act as pressure vessels, preventing over-expansion of the cell when water enters.

The composition of cell walls varies between species and may depend on cell type and developmental stage. The primary cell wall of land plants is composed of the polysaccharides cellulose, hemicellulose and pectin. Often, other polymers such as lignin, suberin or cutin are anchored to or embedded in plant cell walls. Algae possess walls made of glycoproteins and polysaccharides such as carrageenan and agar that are absent from land plants. In bacteria, the cell wall is composed of peptidoglycan. The cell walls of archaea have various compositions, and may be formed of glycoprotein S-layers, pseudopeptidoglycan, or polysaccharides. Fungi possess cell walls made of the glucosamine polymer chitin. Unusually, diatoms have a cell wall composed of biogenic silica.

History

A plant cell wall was first observed and named (simply as a "wall") by Robert Hooke in 1665. However, "the dead excrusion product of the living protoplast" was forgotten, for almost three centuries, being the subject of scientific interest mainly as a resource for industrial processing or in relation to animal or human health.

In 1804, Karl Rudolphi and J.H.F. Link proved that cells had independent cell walls. Before, it had been thought that cells shared walls and that fluid passed between them this way.

The mode of formation of the cell wall was controversial in the 19th century. Hugo von Mohl (1853,

1858) advocated the idea that the cell wall grows by apposition. Carl Nägeli (1858, 1862, 1863) believed that the growth of the wall in thickness and in area was due to a process termed intussusception. Each theory was improved in the following decades: the apposition (or lamination) theory by Eduard Strasburger (1882, 1889), and the intussusception theory by Julius Wiesner (1886).

In 1930, Ernst Münch coined the term apoplast in order to separate the "living" symplast from the "dead" plant region, the latter of which included the cell wall.

By the 1980s, some authors suggested replacing the term "cell wall", particularly as it was used for plants, with the more precise term "extracellular matrix", as used for animal cells, but others preferred the older term.

Properties

Cell walls serve similar purposes in those organisms that possess them. They may give cells rigidity and strength, offering protection against mechanical stress. In multicellular organisms, they permit the organism to build and hold a definite shape (morphogenesis). Cell walls also limit the entry of large molecules that may be toxic to the cell. They further permit the creation of stable osmotic environments by preventing osmotic lysis and helping to retain water. Their composition, properties, and form may change during the cell cycle and depend on growth conditions.

Rigidity of Cell Walls

In most cells, the cell wall is flexible, meaning that it will bend rather than holding a fixed shape, but has considerable tensile strength. The apparent rigidity of primary plant tissues is enabled by cell walls, but is not due to the walls' stiffness. Hydraulic turgor pressure creates this rigidity, along with the wall structure. The flexibility of the cell walls is seen when plants wilt, so that the stems and leaves begin to droop, or in seaweeds that bend in water currents. As John Howland explains:

Think of the cell wall as a wicker basket in which a balloon has been inflated so that it exerts pressure from the inside. Such a basket is very rigid and resistant to mechanical damage. Thus does the prokaryote cell (and eukaryotic cell that possesses a cell wall) gain strength from a flexible plasma membrane pressing against a rigid cell wall.

The apparent rigidity of the cell wall thus results from inflation of the cell contained within. This inflation is a result of the passive uptake of water.

In plants, a secondary cell wall is a thicker additional layer of cellulose which increases wall rigidity. Additional layers may be formed by lignin in xylem cell walls, or suberin in cork cell walls. These compounds are rigid and waterproof, making the secondary wall stiff. Both wood and bark cells of trees have secondary walls. Other parts of plants such as the leaf stalk may acquire similar reinforcement to resist the strain of physical forces.

Permeability

The primary cell wall of most plant cells is freely permeable to small molecules including small proteins, with size exclusion estimated to be 30-60 kDa. The pH is an important factor governing the transport of molecules through cell walls.

Evolution

Cell walls evolved independently in many groups, even in the photosynthetic eukaryotes. In these lineages, the cell wall is closely related to the evolution of multicellularity, terrestrialization and vascularization.

Plant Cell Walls

The walls of plant cells must have sufficient tensile strength to withstand internal osmotic pressures of several times atmospheric pressure that result from the difference in solute concentration between the cell interior and external solutions. Plant cell walls vary from 0.1 to several μm in thickness.

Layers

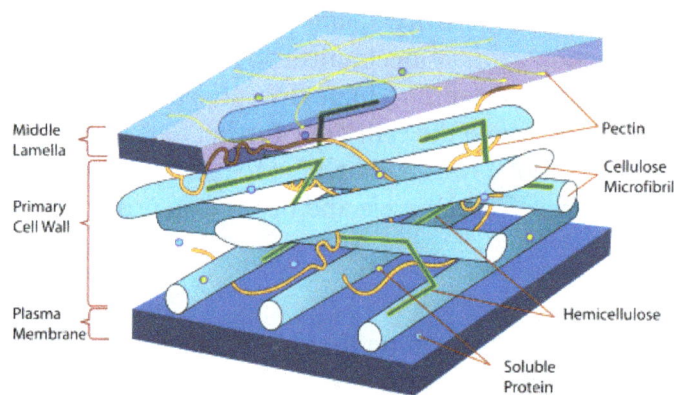

Molecular structure of the primary cell wall in plants.

Up to three strata or layers may be found in plant cell walls:

- The primary cell wall, generally a thin, flexible and extensible layer formed while the cell is growing.

- The secondary cell wall, a thick layer formed inside the primary cell wall after the cell is fully grown. It is not found in all cell types. Some cells, such as the conducting cells in xylem, possess a secondary wall containing lignin, which strengthens and waterproofs the wall.

- The middle lamella, a layer rich in pectins. This outermost layer forms the interface between adjacent plant cells and glues them together.

Composition

In the primary (growing) plant cell wall, the major carbohydrates are cellulose, hemicellulose and pectin. The cellulose microfibrils are linked via hemicellulosic tethers to form the cellulose-hemicellulose network, which is embedded in the pectin matrix. The most common hemicellulose in the primary cell wall is xyloglucan. In grass cell walls, xyloglucan and pectin are reduced in abundance and partially replaced by glucuronarabinoxylan, another type of hemicellulose. Primary cell walls characteristically extend (grow) by a mechanism called acid growth, which involves turgor-driven movement of the strong cellulose microfibrils within the weaker hemicellulose/pectin matrix, catalyzed by expansin proteins. The outer part of the primary cell wall of the plant epidermis is usually impregnated with cutin and wax, forming a permeability barrier known as the plant cuticle.

Secondary cell walls contain a wide range of additional compounds that modify their mechanical properties and permeability. The major polymers that make up wood (largely secondary cell walls) include:

- cellulose, 35-50%

- xylan, 20-35%, a type of hemicellulose

- lignin, 10-25%, a complex phenolic polymer that penetrates the spaces in the cell wall between cellulose, hemicellulose and pectin components, driving out water and strengthening the wall.

Additionally, structural proteins (1-5%) are found in most plant cell walls; they are classified as hydroxyproline-rich glycoproteins (HRGP), arabinogalactan proteins (AGP), glycine-rich proteins (GRPs), and proline-rich proteins (PRPs). Each class of glycoprotein is defined by a characteristic, highly repetitive protein sequence. Most are glycosylated, contain hydroxyproline (Hyp) and become cross-linked in the cell wall. These proteins are often concentrated in specialized cells and in cell corners. Cell walls of the epidermis may contain cutin. The Casparian strip in the endodermis roots and cork cells of plant bark contain suberin. Both cutin and suberin are polyesters that function as permeability barriers to the movement of water. The relative composition of carbohydrates, secondary compounds and proteins varies between plants and between the cell type and age. Plant cells walls also contain numerous enzymes, such as hydrolases, esterases, peroxidases, and transglycosylases, that cut, trim and cross-link wall polymers.

Secondary walls - especially in grasses - may also contain microscopic silica crystals, which may strengthen the wall and protect it from herbivores.

Cell walls in some plant tissues also function as storage deposits for carbohydrates that can be broken down and resorbed to supply the metabolic and growth needs of the plant. For example, endosperm cell walls in the seeds of cereal grasses, nasturtium and other species, are rich in glucans and other polysaccharides that are readily digested by enzymes during seed germination to form simple sugars that nourish the growing embryo.

Formation

The middle lamella is laid down first, formed from the cell plate during cytokinesis, and the primary cell wall is then deposited inside the middle lamella. The actual structure of the cell wall is not clearly defined and several models exist - the covalently linked cross model, the tether model, the diffuse layer model and the stratified layer model. However, the primary cell wall, can be defined as composed of cellulose microfibrils aligned at all angles. Cellulose microfibrils are produced at the plasma membrane by the cellulose synthase complex, which is proposed to be made of a hexameric rosette that contains three cellulose synthase catalytic subunits for each of the six units. Microfibrils are held together by hydrogen bonds to provide a high tensile strength. The cells are held together and share the gelatinous membrane called the *middle lamella*, which contains magnesium and calcium pectates (salts of pectic acid). Cells interact though plasmodesmata, which are inter-connecting channels of cytoplasm that connect to the protoplasts of adjacent cells across the cell wall.

In some plants and cell types, after a maximum size or point in development has been reached, a *secondary wall* is constructed between the plasma membrane and primary wall. Unlike the primary wall, the cellulose microfibrils are aligned parallel in layers, the orientation changing slightly with each additional layer so that the structure becomes helicoidal. Cells with secondary cell walls can be rigid, as in the gritty sclereid cells in pear and quince fruit. Cell to cell communication is possible through pits in the secondary cell wall that allow plasmodesmata to connect cells through the secondary cell walls.

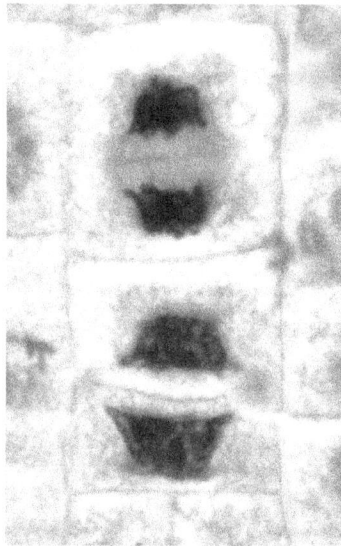

Photomicrograph of onion root cells, showing the centrifugal development of new cell walls (phragmoplast).

Fungal Cell Walls

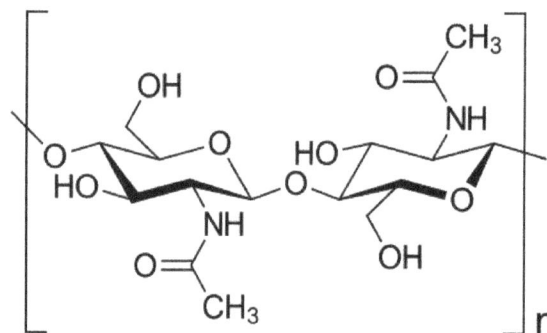

Chemical structure of a unit from a chitin polymer chain.

There are several groups of organisms that have been called "fungi". Some of these groups (Oomycete and Myxogastria) have been transferred out of the Kingdom Fungi, in part because of fundamental biochemical differences in the composition of the cell wall. Most true fungi have a cell wall consisting largely of chitin and other polysaccharides. True fungi do not have cellulose in their cell walls.

True Fungi

In fungi, the cell wall is the outer-most layer, external to the plasma membrane. The fungal cell wall is a matrix of three main components:

- chitin: polymers consisting mainly of unbranched chains of β-(1,4)-linked-N-Acetylglucos-amine in the Ascomycota and Basidiomycota, or poly-β-(1,4)-linked-N-Acetylglucosamine (chitosan) in the Zygomycota. Both chitin and chitosan are synthesized and extruded at the plasma membrane.

- glucans: glucose polymers that function to cross-link chitin or chitosan polymers. β-glucans are glucose molecules linked via β-(1,3)- or β-(1,6)- bonds and provide rigidity to the cell wall while α-glucans are defined by α-(1,3)- and/or α-(1,4) bonds and function as part of the matrix.

- proteins: enzymes necessary for cell wall synthesis and lysis in addition to structural proteins are all present in the cell wall. Most of the structural proteins found in the cell wall are glycosylated and contain mannose, thus these proteins are called mannoproteins or mannans.

Other Eukaryotic Cell Walls

Algae

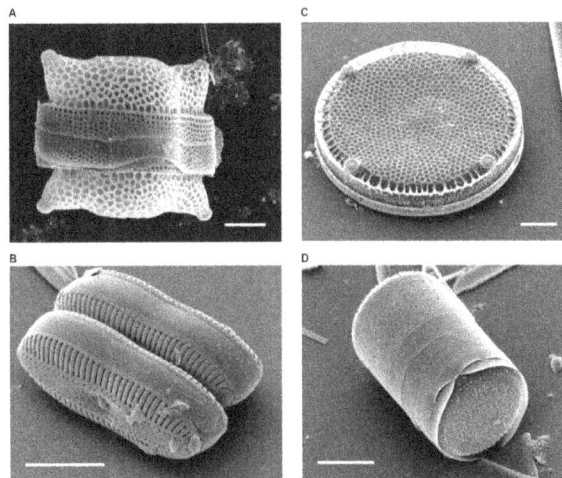

Scanning electron micrographs of diatoms showing
the external appearance of the cell wall

Like plants, algae have cell walls. Algal cell walls contain either polysaccharides (such as cellulose (a glucan)) or a variety of glycoproteins (Volvocales) or both. The inclusion of additional polysaccharides in algal cells walls is used as a feature for algal taxonomy.

- Mannans: They form microfibrils in the cell walls of a number of marine green algae including those from the genera, *Codium*, *Dasycladus*, and *Acetabularia* as well as in the walls of some red algae, like *Porphyra* and *Bangia*.

- Xylans:

- Alginic acid: It is a common polysaccharide in the cell walls of brown algae.

- Sulfonated polysaccharides: They occur in the cell walls of most algae; those common in red algae include agarose, carrageenan, porphyran, furcelleran and funoran.

Other compounds that may accumulate in algal cell walls include sporopollenin and calcium ions.

The group of algae known as the diatoms synthesize their cell walls (also known as frustules or valves) from silicic acid (specifically orthosilicic acid, H_4SiO_4). The acid is polymerised intra-cellularly, then the wall is extruded to protect the cell. Significantly, relative to the organic cell walls produced by other groups, silica frustules require less energy to synthesize (approximately 8%), potentially a major saving on the overall cell energy budget and possibly an explanation for higher growth rates in diatoms.

In brown algae, phlorotannins may be a constituent of the cell walls.

Water Molds

The group Oomycetes, also known as water molds, are saprotrophic plant pathogens like fungi. Until recently they were widely believed to be fungi, but structural and molecular evidence has led to their reclassification as heterokonts, related to autotrophic brown algae and diatoms. Unlike fungi, oomycetes typically possess cell walls of cellulose and glucans rather than chitin, although some genera (such as *Achlya* and *Saprolegnia*) do have chitin in their walls. The fraction of cellulose in the walls is no more than 4 to 20%, far less than the fraction of glucans. Oomycete cell walls also contain the amino acid hydroxyproline, which is not found in fungal cell walls.

Slime Molds

The dictyostelids are another group formerly classified among the fungi. They are slime molds that feed as unicellular amoebae, but aggregate into a reproductive stalk and sporangium under certain conditions. Cells of the reproductive stalk, as well as the spores formed at the apex, possess a cellulose wall. The spore wall has been shown to possess three layers, the middle of which is composed primarily of cellulose, and the innermost is sensitive to cellulase and pronase.

Prokaryotic Cell Walls

Bacterial Cell Walls

Around the outside of the cell membrane is the bacterial cell wall. Bacterial cell walls are made of peptidoglycan (also called murein), which is made from polysaccharide chains cross-linked by unusual peptides containing D-amino acids. Bacterial cell walls are different from the cell walls of plants and fungi which are made of cellulose and chitin, respectively. The cell wall of bacteria is also distinct from that of Archaea, which do not contain peptidoglycan. The cell wall is essential to the survival of many bacteria, although L-form bacteria can be produced in the laboratory that lack a cell wall. The antibiotic penicillin is able to kill bacteria by preventing the cross-linking of peptidoglycan and this causes the cell wall to weaken and lyse. The lysozyme enzyme can also damage bacterial cell walls.

There are broadly speaking two different types of cell wall in bacteria, called Gram-positive and Gram-negative. The names originate from the reaction of cells to the Gram stain, a test long-employed for the classification of bacterial species.

Gram-positive bacteria possess a thick cell wall containing many layers of peptidoglycan and teichoic acids. In contrast, Gram-negative bacteria have a relatively thin cell wall consisting of a few layers of peptidoglycan surrounded by a second lipid membrane containing lipopolysaccharides and lipoproteins. Most bacteria have the Gram-negative cell wall and only the Firmicutes and Actinobacteria (previously known as the low G+C and high G+C Gram-positive bacteria, respectively) have the alternative Gram-positive arrangement. These differences in structure can produce differences in antibiotic susceptibility, for instance vancomycin can kill only Gram-positive bacteria and is ineffective against Gram-negative pathogens, such as *Haemophilus influenzae* or *Pseudomonas aeruginosa*.

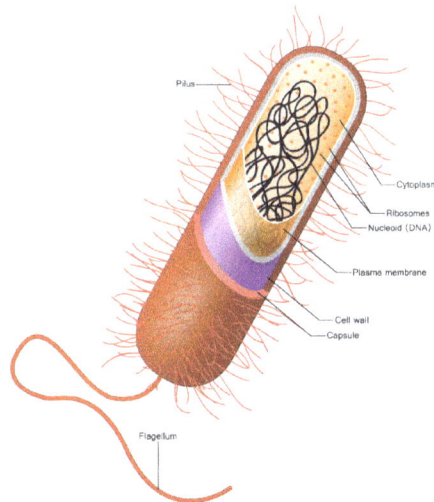

Diagram of a typical Gram-positive bacterium. The cell envelope comprises a plasma membrane, seen here in light brown, and a thick peptidoglycan-containing cell wall (the purple layer).
No outer lipid membrane is present, as would be
the case in Gram-negative bacteria. The red layer, known as the capsule, is distinct from the cell envelope.

Archaeal Cell Walls

Although not truly unique, the cell walls of Archaea are unusual. Whereas peptidoglycan is a standard component of all bacterial cell walls, all archaeal cell walls lack peptidoglycan, with the exception of one group of methanogens. In that group, the peptidoglycan is a modified form very different from the kind found in bacteria. There are four types of cell wall currently known among the Archaea.

One type of archaeal cell wall is that composed of pseudopeptidoglycan (also called pseudomurein). This type of wall is found in some methanogens, such as *Methanobacterium* and *Methanothermus*. While the overall structure of archaeal *pseudo*peptidoglycan superficially resembles that of bacterial peptidoglycan, there are a number of significant chemical differences. Like the peptidoglycan found in bacterial cell walls, pseudopeptidoglycan consists of polymer chains of glycan cross-linked by short peptide connections. However, unlike peptidoglycan, the sugar N-acetylmuramic acid is replaced by N-acetyltalosaminuronic acid, and the two sugars are bonded with a

β,1-3 glycosidic linkage instead of β,1-4. Additionally, the cross-linking peptides are L-amino acids rather than D-amino acids as they are in bacteria.

A second type of archaeal cell wall is found in *Methanosarcina* and *Halococcus*. This type of cell wall is composed entirely of a thick layer of polysaccharides, which may be sulfated in the case of *Halococcus*. Structure in this type of wall is complex and not fully investigated.

A third type of wall among the Archaea consists of glycoprotein, and occurs in the hyperthermophiles, *Halobacterium*, and some methanogens. In *Halobacterium*, the proteins in the wall have a high content of acidic amino acids, giving the wall an overall negative charge. The result is an unstable structure that is stabilized by the presence of large quantities of positive sodium ions that neutralize the charge. Consequently, *Halobacterium* thrives only under conditions with high salinity.

In other Archaea, such as *Methanomicrobium* and *Desulfurococcus*, the wall may be composed only of surface-layer proteins, known as an *S-layer*. S-layers are common in bacteria, where they serve as either the sole cell-wall component or an outer layer in conjunction with polysaccharides. Most Archaea are Gram-negative, though at least one Gram-positive member is known.

Other Cell Coverings

Many protists and bacteria produce other cell surface structures apart from cell walls, external (extracellular matrix) or internal. Many algae have a sheath or envelope of mucilage outside the cell made of exopolysaccharides. Diatoms build a frustule from silica extracted from the surrounding water; radiolarians, foraminiferans, testate amoebae and silicoflagellates also produce a skeleton from minerals, called test in some groups. Many green algae, such as *Halimeda* and the Dasycladales, and some red algae, the Corallinales, encase their cells in a secreted skeleton of calcium carbonate. In each case, the wall is rigid and essentially inorganic. It is the non-living component of cell. Some golden algae, ciliates and choanoflagellates produces a shell-like protective outer covering called lorica. Some dinoflagellates have a theca of cellulose plates, and coccolithophorids have coccoliths.

An extracellular matrix is also present in metazoans. Its composition varies between cells, but collagens are the most abundant protein in the ECM.

Hemicellulose

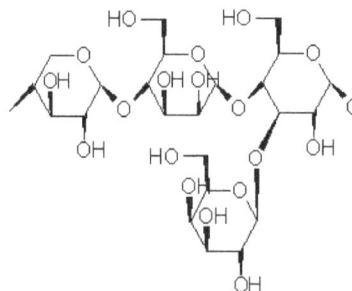

- Xylose - ß(1,4) - Mannose - ß(1,4) - Glucose -
- alpha(1,3) - Galactose

Hemicellulose

Most common molecular motif of hemicellulose

A hemicellulose (also known as polyose) is any of several heteropolymers (matrix polysaccharides), such as arabinoxylans, present along with cellulose in almost all plant cell walls. While cellulose is crystalline, strong, and resistant to hydrolysis, hemicellulose has a random, amorphous structure with little strength. It is easily hydrolyzed by dilute acid or base as well as myriad hemicellulase enzymes.

Composition

Hemicelluloses include xylan, glucuronoxylan, arabinoxylan, glucomannan, and xyloglucan.

These polysaccharides contain many different sugar monomers. In contrast, cellulose contains only anhydrous glucose. For instance, besides glucose, sugar monomers in hemicellulose can include xylose, mannose, galactose, rhamnose, and arabinose. Hemicelluloses contain most of the D-pentose sugars, and occasionally small amounts of L-sugars as well. Xylose is in most cases the sugar monomer present in the largest amount, although in softwoods mannose can be the most abundant sugar. Not only regular sugars can be found in hemicellulose, but also their acidified form, for instance glucuronic acid and galacturonic acid can be present. Hemicellulose is often associated with cellulose, but it has different composition.

Structural Comparison to Cellulose

Unlike cellulose, hemicellulose (also a polysaccharide) consists of shorter chains – 500–3,000 sugar units as opposed to 7,000–15,000 glucose molecules per polymer, as seen in cellulose. In addition, hemicellulose is a branched polymer, while cellulose is unbranched.

Native Structure

Hemicelluloses are embedded in the cell walls of plants, sometimes in chains that form a 'ground' - they bind with pectin to cellulose to form a network of cross-linked fibres.

Biosynthesis

Hemicelluloses are synthesised from sugar nucleotides in the cell's Golgi apparatus. Two models explain their synthesis: 1) a '2 component model' where modification occurs at two transmembrane proteins, and 2) a '1 component model' where modification occurs only at one transmembrane protein. After synthesis, hemicelluloses are transported to the plasma membrane via Golgi vesicles.

Applications

As percent content of hemicellulose increases in animal feed, the voluntary feed intake decreases.

Hemicellulose is represented by the difference between neutral detergent fiber (NDF) and acid detergent fiber (ADF).

Functions

Microfibrils are cross-linked together by hemicellulose homopolymers. Lignins assist and strengthen the attachment of hemicelluloses to microfibrils.

Hemicellulose from Trees

Hemicellulose found in hardwood trees is predominantly xylan with some glucomannan, while in softwoods it is mainly rich in galactoglucomannan and contains only a small amount of xylan. The average molecular weight is lower than that of cellulose at less than 30,000, as opposed to the 100,000 average molecular weight reported for cellulose.

Vacuole

A vacuole is a membrane-bound organelle which is present in all plant and fungal cells and some protist, animal and bacterial cells. Vacuoles are essentially enclosed compartments which are filled with water containing inorganic and organic molecules including enzymes in solution, though in certain cases they may contain solids which have been engulfed. Vacuoles are formed by the fusion of multiple membrane vesicles and are effectively just larger forms of these. The organelle has no basic shape or size; its structure varies according to the needs of the cell.

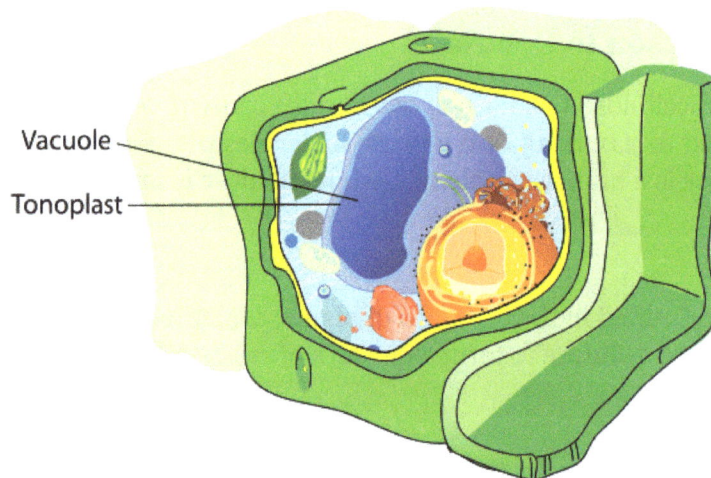

Plant cell structure

The function and significance of vacuoles varies greatly according to the type of cell in which they are present, having much greater prominence in the cells of plants, fungi and certain protists than those of animals and bacteria. In general, the functions of the vacuole include:

- Isolating materials that might be harmful or a threat to the cell

- Containing waste products

- Containing water in plant cells

- Maintaining internal hydrostatic pressure or turgor within the cell

- Maintaining an acidic internal pH

- Containing small molecules

- Exporting unwanted substances from the cell

- Allows plants to support structures such as leaves and flowers due to the pressure of the central vacuole

- By increasing in size, allows the germinating plant or its organs (such as leaves) to grow very quickly and using up mostly just water.

- In seeds, stored proteins needed for germination are kept in 'protein bodies', which are modified vacuoles.

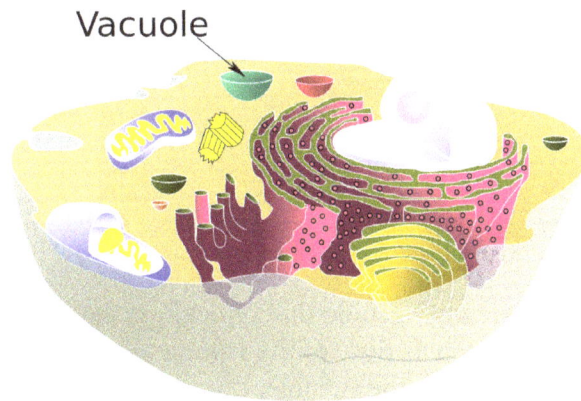

Animal cell structure

Vacuoles also play a major role in autophagy, maintaining a balance between biogenesis (production) and degradation (or turnover), of many substances and cell structures in certain organisms. They also aid in the lysis and recycling of misfolded proteins that have begun to build up within the cell. Thomas Boller and others proposed that the vacuole participates in the destruction of invading bacteria and Robert B Mellor proposed organ-specific forms have a role in 'housing' symbiotic bacteria. In protists, vacuoles have the additional function of storing food which has been absorbed by the organism and assisting in the digestive and waste management process for the cell.

The vacuole probably evolved several times independently, even within the Viridiplantae.

Discovery

Contractile vacuoles ("stars") were first observed by Spallanzani (1776) in protozoa, although mistaken for respiratory organs. Dujardin (1841) named these "stars" as *vacuoles*. In 1842, Schleiden applied the term for plant cells, to distinguish the structure with cell sap from the rest of the protoplasm.

In 1885, de Vries named the vacuoule membrane as tonoplast.

Bacteria

Large vacuoles are found in three genera of filamentous sulfur bacteria, the *Thioploca*, *Beggiatoa* and *Thiomargarita*. The cytosol is extremely reduced in these genera and the vacuole can occupy between 40–98% of the cell. The vacuole contains high concentrations of nitrate ions and is therefore thought to be a storage organelle.

Gas vacuoles, which are freely permeable to gas, are present in some species of Cyanobacteria. They allow the bacteria to control their buoyancy.

Plants

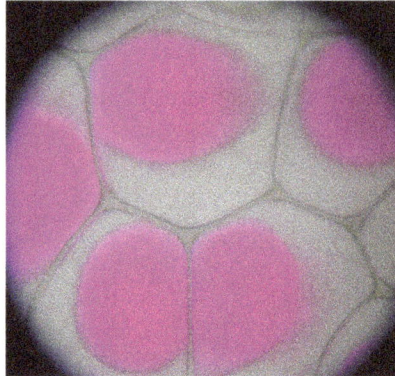

The anthocyanin-storing vacuoles of *Rhoeo spathacea*,
a spiderwort, in cells that have plasmolyzed

Most mature plant cells have one large vacuole that typically occupies more than 30% of the cell's volume, and that can occupy as much as 80% of the volume for certain cell types and conditions. Strands of cytoplasm often run through the vacuole.

A vacuole is surrounded by a membrane called the tonoplast (word origin: Gk tón(os) + -o-, meaning "stretching", "tension", "tone" + comb. form repr. Gk plastós formed, molded) and filled with cell sap. Also called the vacuolar membrane, the tonoplast is the cytoplasmic membrane surrounding a vacuole, separating the vacuolar contents from the cell's cytoplasm. As a membrane, it is mainly involved in regulating the movements of ions around the cell, and isolating materials that might be harmful or a threat to the cell.

Transport of protons from the cytosol to the vacuole stabilizes cytoplasmic pH, while making the vacuolar interior more acidic creating a proton motive force which the cell can use to transport nutrients into or out of the vacuole. The low pH of the vacuole also allows degradative enzymes to act. Although single large vacuoles are most common, the size and number of vacuoles may vary in different tissues and stages of development. For example, developing cells in the meristems contain small provacuoles and cells of the vascular cambium have many small vacuoles in the winter and one large one in the summer.

Aside from storage, the main role of the central vacuole is to maintain turgor pressure against the cell wall. Proteins found in the tonoplast (aquaporins) control the flow of water into and out of the vacuole through active transport, pumping potassium (K^+) ions into and out of the vacuolar interior. Due to osmosis, water will diffuse into the vacuole, placing pressure on the cell wall. If water loss leads to a significant decline in turgor pressure, the cell will plasmolyze. Turgor pressure exerted by vacuoles is also required for cellular elongation: as the cell wall is partially degraded by the action of expansins, the less rigid wall is expanded by the pressure coming from within the vacuole. Turgor pressure exerted by the vacuole is also essential in supporting plants in an upright position. Another function of a central vacuole is that it pushes all contents of the cell's cytoplasm against the cellular membrane, and thus keeps the chloroplasts closer to light. Most plants store chemicals in the vacuole that react with chemicals in the cytosol. If the cell is broken, for example by a herbivore, then the two chemicals can react forming toxic chemicals. In garlic, alliin and the enzyme alliinase are normally separated but form allicin if the vacuole is broken. A similar reaction is responsible for the production of syn-propanethial-S-oxide when onions are cut.

Fungi

Vacuoles in fungal cells perform similar functions to those in plants and there can be more than one vacuole per cell. In yeast cells the vacuole is a dynamic structure that can rapidly modify its morphology. They are involved in many processes including the homeostasis of cell pH and the concentration of ions, osmoregulation, storing amino acids and polyphosphate and degradative processes. Toxic ions, such as strontium (Sr^{2+}), cobalt(II) (Co^{2+}), and lead(II) (Pb^{2+}) are transported into the vacuole to isolate them from the rest of the cell.

Animals

In animal cells, vacuoles perform mostly subordinate roles, assisting in larger processes of exocytosis and endocytosis.

Animal vacuoles are smaller than their plant counterparts but also usually greater in number. There are also animal cells that do not have any vacuoles.

Exocytosis is the extrusion process of proteins and lipids from the cell. These materials are absorbed into secretory granules within the Golgi apparatus before being transported to the cell membrane and secreted into the extracellular environment. In this capacity, vacuoles are simply storage vesicles which allow for the containment, transport and disposal of selected proteins and lipids to the extracellular environment of the cell.

Endocytosis is the reverse of exocytosis and can occur in a variety of forms. Phagocytosis ("cell eating") is the process by which bacteria, dead tissue, or other bits of material visible under the microscope are engulfed by cells. The material makes contact with the cell membrane, which then invaginates. The invagination is pinched off, leaving the engulfed material in the membrane-enclosed vacuole and the cell membrane intact. Pinocytosis ("cell drinking") is essentially the same process, the difference being that the substances ingested are in solution and not visible under the microscope. Phagocytosis and pinocytosis are both undertaken in association with lysosomes which complete the breakdown of the material which has been engulfed.

Salmonella is able to survive and reproduce in the vacuoles of several mammal species after being engulfed.

References

- Reggiori F (2006). "Membrane Origin for Autophagy". Current Topics in Developmental Biology. 74: 1–30. PMID 16860663. doi:10.1016/S0070-2153(06)74001-7

- Raven, PH; Evert, RF; Eichhorm, SE (1999). Biology of Plants (6th ed.). New York: W.H. Freeman. ISBN 9780716762843

- Lewis, LA; McCourt, RM (2004). "Green algae and the origin of land plants". American Journal of Botany. 91: 1535–1556. PMID 21652308. doi:10.3732/ajb.91.10.1535

- Kalenderblatt Dezember 2013 – Mathematisch-Naturwissenschaftliche Fakultät – Universität Rostock. Math-nat.uni-rostock.de (2013-11-28). Retrieved on 2015-10-15

- Raven, JA (1997). "The vacuole: a cost-benefit analysis". Advances in Botanical Research. 25: 59–86. doi:10.1016/S0065-2296(08)60148-2

- Howland, John L. (2000). The Surprising Archaea: Discovering Another Domain of Life. Oxford: Oxford University Press. pp. 69–71. ISBN 0-19-511183-4

- Manton, I; Clarke, B (1952). "An electron microscope study of the spermatozoid of Sphagnum". Journal of Experimental Botany. 3: 265–275. doi:10.1093/jxb/3.3.265

- Sendbusch, Peter V. (2003-07-31). "Cell Walls of Algae Archived November 28, 2005, at the Wayback Machine.". Botany Online. Retrieved on 2007-10-29

- Paolillo, Jr., DJ (1967). "On the structure of the axoneme in flagella of Polytrichum juniperinum". Transactions of the American Microscopical Society. 86: 428–433. doi:10.2307/3224266

- Buchanan; Gruissem, Jones (2000). Biochemistry & molecular biology of plants (1st ed.). American society of plant physiology. ISBN 0-943088-39-9

- Gibson LJ (2013). "The hierarchical structure and mechanics of plant materials". Journal of the Royal Society Interface. 9 (76): 2749–2766. PMC 3479918. PMID 22874093. doi:10.1098/rsif.2012.0341

- Heide N. Schulz-Vogt (2006). Vacuoles. Microbiology Monographs. 1. ISBN 3-540-26205-9. doi:10.1007/3-540-33774-1_10

- Brock, Thomas D., Michael T. Madigan, John M. Martinko, & Jack Parker. (1994) Biology of Microorganisms, 7th ed., pages 818-819, 824 (Englewood Cliffs, NJ: Prentice Hall). ISBN 0-13-042169-3

- Sengbusch, Peter V. (2003-07-31). "Interactions between Plants and Fungi: the Evolution of their Parasitic and Symbiotic Relations Archived December 8, 2006, at the Wayback Machine.". biologie.uni-hamburg.de. Retrieved on 2007-10-29

- Furnas, M. J. (1990). "In situ growth rates of marine phytoplankton : Approaches to measurement, community and species growth rates". J. Plankton Res. 12 (6): 1117–1151. doi:10.1093/plankt/12.6.1117

An Integrated Study of Plastid

Plastids are double membrane organelle found in plants and selected algae. Chloroplast, chromoplast, leucoplast and gerontoplast are some categories of plastids. They have the ability to change into other cells depending on the function they play. The major categories of plastids are dealt with great details in the chapter.

Plastid

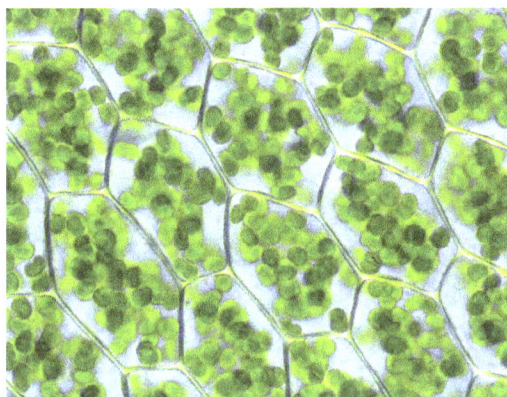

Plant cells with visible chloroplasts.

The plastid (Greek plastós: formed, molded – plural plastids) is a major double-membrane organelle found, among others, in the cells of plants and algae. Plastids are the site of manufacture and storage of important chemical compounds used by the cell. They often contain pigments used in photosynthesis, and the types of pigments present can change or determine the cell's color. They have a common evolutionary origin and possess a double-stranded DNA molecule that is circular, like that of prokaryotic cells.

Plastids in Plants

Those plastids that contain chlorophyll can carry out photosynthesis. Plastids can also store products like starch and can synthesize fatty acids and terpenes, which can be used for producing energy and as raw material for the synthesis of other molecules. For example, the components of the plant cuticle and its epicuticular wax are synthesized by the epidermal cells from palmitic acid, which is synthesized in the chloroplasts of the mesophyll tissue. All plastids are derived from proplastids, which are present in the meristematic regions of the plant. Proplastids and young chloroplasts commonly divide by binary fission, but more mature chloroplasts also have this capacity.

- Chloroplasts green plastids: for photosynthesis; *the predecessors of chloroplasts*

- Chromoplasts coloured plastids: for pigment synthesis and storage

- Gerontoplasts: control the dismantling of the photosynthetic apparatus during plant senescence

- Leucoplasts colourless plastids: for monoterpene synthesis; *leucoplasts sometimes differentiate into more specialized plastids:*

- Amyloplasts: for starch storage and detecting gravity

- Elaioplasts: for storing fat

- Proteinoplasts: for storing and modifying protein

- Tannosomes: for synthesizing and producing tannins and polyphenols

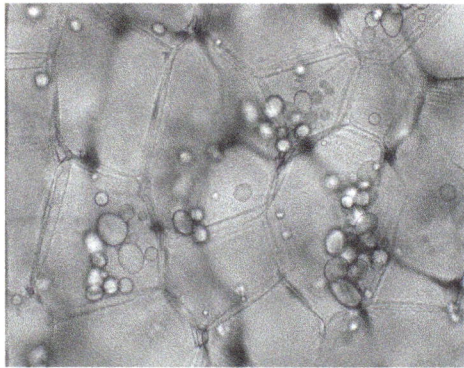

Leucoplasts in plant cells.

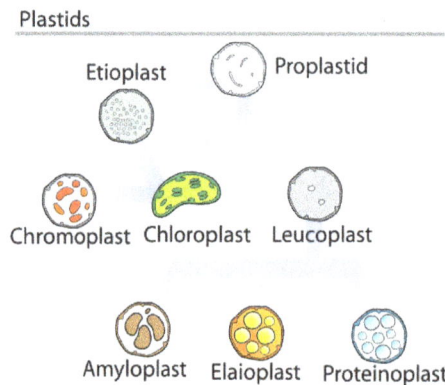

In plants, plastids may differentiate into several forms, depending upon which function they play in the cell. Undifferentiated plastids (*proplastids*) may develop into any of the following variants:

Depending on their morphology and function, plastids have the ability to differentiate, or redifferentiate, between these and other forms.

Each plastid creates multiple copies of a circular 75–250 kilobase plastome. The number of genome copies per plastid is variable, ranging from more than 1000 in rapidly dividing cells, which, in general, contain few plastids, to 100 or fewer in mature cells, where plastid divisions have given rise to a large number of plastids. The plastome contains about 100 genes encoding ribosomal and transfer ribonucleic acids (rRNAs and tRNAs) as well as proteins involved in photosynthesis and plastid gene transcription and translation. However, these proteins only represent a small fraction of the total protein set-up necessary to build and maintain the structure and function of a particular type of plastid. Plant nuclear genes encode the vast majority of plastid proteins, and the expression of plastid genes and nuclear genes is tightly co-regulated to coordinate proper development of plastids in relation to cell differentiation.

Plastid DNA exists as large protein-DNA complexes associated with the inner envelope membrane and called 'plastid nucleoids'. Each nucleoid particle may contain more than 10 copies of the plastid DNA. The proplastid contains a single nucleoid located in the centre of the plastid. The developing plastid has many nucleoids, localized at the periphery of the plastid, bound to the inner envelope membrane. During the development of proplastids to chloroplasts, and when plastids convert from one type to another, nucleoids change in morphology, size and location within the organelle. The remodelling of nucleoids is believed to occur by modifications to the composition and abundance of nucleoid proteins.

Many plastids, particularly those responsible for photosynthesis, possess numerous internal membrane layers.

In plant cells, long thin protuberances called stromules sometimes form and extend from the main plastid body into the cytosol and interconnect several plastids. Proteins, and presumably smaller molecules, can move within stromules. Most cultured cells that are relatively large compared to other plant cells have very long and abundant stromules that extend to the cell periphery.

In 2014, evidence of possible plastid genome loss was found in *Rafflesia lagascae,* a non-photosynthetic parasitic flowering plant, and in *Polytomella,* a genus of non-photosynthetic green algae. Extensive searches for plastid genes in both *Rafflesia* and *Polytomella* yielded no results, however the conclusion that their plastomes are entirely missing is still controversial. Some scientists argue that plastid genome loss is unlikely since even non-photosynthetic plastids contain genes necessary to complete various biosynthetic pathways, such as heme biosynthesis.

Plastids in Algae

In algae, the term leucoplast is used for all unpigmented plastids and their function differs from the leucoplasts of plants. Etioplasts, amyloplasts and chromoplasts are plant-specific and do not occur in algae. Plastids in algae and hornworts may also differ from plant plastids in that they contain pyrenoids.

Glaucocystophytic algae contain muroplasts, which are similar to chloroplasts except that they have a peptidoglycan cell wall that is similar to that of prokaryotes. Rhodophytic algae contain rhodoplasts, which are red chloroplasts that allow the algae to photosynthesise to a depth of up to 268 m. The chloroplasts of plants differ from the rhodoplasts of red algae in their ability to synthesize starch, which is stored in the form of granules within the plastids. In red algae, floridean starch is synthesized and stored outside the plastids in the cytosol.

Inheritance of Plastids

Most plants inherit the plastids from only one parent. In general, angiosperms inherit plastids from the female gamete, whereas many gymnosperms inherit plastids from the male pollen.

Algae also inherit plastids from only one parent. The plastid DNA of the other parent is, thus, completely lost.

In normal intraspecific crossings (resulting in normal hybrids of one species), the inheritance of plastid DNA appears to be quite strictly 100% uniparental. In interspecific hybridisations, however, the inheritance of plastids appears to be more erratic. Although plastids inherit mainly maternally in interspecific hybridisations, there are many reports of hybrids of flowering plants that contain plastids of the father. Approximately 20% of angiosperms, including alfalfa (*Medicago sativa*), normally show biparental inheritance of plastids.

Origin of Plastids

Plastids are thought to have originated from endosymbiotic cyanobacteria. This symbiosis evolved around 1.5 billion years ago and enabled eukaryotes to carry out oxygenic photosynthesis. Three evolutionary lineages have since emerged in which the plastids are named differently: chloroplasts in green algae and plants, rhodoplasts in red algae and muroplasts in the glaucophytes. The plastids differ both in their pigmentation and in their ultrastructure. For example, chloroplasts have lost all phycobilisomes, the light harvesting complexes found in cyanobacteria, red algae and glaucophytes, but instead contain stroma and grana thylakoids, structures found only in plants and closely related green algae. The glaucocystophycean plastid — in contrast to chloroplasts and rhodoplasts — is still surrounded by the remains of the cyanobacterial cell wall. All these primary plastids are surrounded by two membranes.

Complex plastids start by secondary endosymbiosis (where a eukaryotic organism engulfs another eukaryotic organism that contains a primary plastid resulting in its endosymbiotic fixation), when a eukaryote engulfs a red or green alga and retains the algal plastid, which is typically surrounded by more than two membranes. In some cases these plastids may be reduced in their metabolic and/or photosynthetic capacity. Algae with complex plastids derived by secondary endosymbiosis of a red alga include the heterokonts, haptophytes, cryptomonads, and most dinoflagellates (= rhodoplasts). Those that endosymbiosed a green alga include the euglenids and chlorarachniophytes (= chloroplasts). The Apicomplexa, a phylum of obligate parasitic protozoa including the causative agents of malaria (*Plasmodium* spp.), toxoplasmosis (*Toxoplasma gondii*), and many other human or animal diseases also harbor a complex plastid (although this organelle has been lost in some apicomplexans, such as *Cryptosporidium parvum*, which causes cryptosporidiosis). The 'apicoplast' is no longer capable of photosynthesis, but is an essential organelle, and a promising target for antiparasitic drug development.

Some dinoflagellates and sea slugs, in particular of the genus *Elysia*, take up algae as food and keep the plastid of the digested alga to profit from the photosynthesis; after a while, the plastids are also digested. This process is known as kleptoplasty, from the Greek, *kleptes*, thief.

Chloroplast

Chloroplasts are organelles, specialized subunits, in plant and algal cells. Their discovery inside plant cells is usually credited to Julius von Sachs (1832–1897), an influential botanist and author of standard botanical textbooks – sometimes called "The Father of Plant Physiology".

The main role of chloroplasts is to conduct photosynthesis, where the photosynthetic pigment chlorophyll captures the energy from sunlight and converts it and stores it in the energy-storage molecules ATP and NADPH while freeing oxygen from water. They then use the ATP and NADPH to make organic molecules from carbon dioxide in a process known as the Calvin cycle. Chloroplasts carry out a number of other functions, including fatty acid synthesis, much amino acid synthesis, and the immune response in plants. The number of chloroplasts per cell varies from one, in unicellular algae, up to 100 in plants like *Arabidopsis* and wheat.

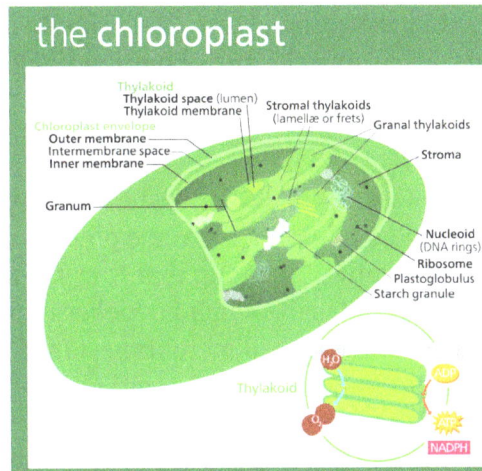

Structure of a typical higher-plant chloroplast

A chloroplast is a type of organelle known as a plastid, characterized by its high concentration of chlorophyll. Other plastid types, such as the leucoplast and the chromoplast, contain little chlorophyll and do not carry out photosynthesis.

Chloroplasts are highly dynamic—they circulate and are moved around within plant cells, and occasionally pinch in two to reproduce. Their behavior is strongly influenced by environmental factors like light color and intensity. Chloroplasts, like mitochondria, contain their own DNA, which is thought to be inherited from their ancestor—a photosynthetic cyanobacterium that was engulfed by an early eukaryotic cell. Chloroplasts cannot be made by the plant cell and must be inherited by each daughter cell during cell division.

With one exception (the amoeboid *Paulinella chromatophora*), all chloroplasts can probably be traced back to a single endosymbiotic event, when a cyanobacterium was engulfed by the eukaryote. Despite this, chloroplasts can be found in an extremely wide set of organisms, some not even directly related to each other—a consequence of many secondary and even tertiary endosymbiotic events.

The word *chloroplast* is derived from the Greek words *chloros*, which means green, and *plastes*, which means "the one who forms".

Discovery

The first definitive description of a chloroplast (*Chlorophyllkörnen*, "grain of chlorophyll") was given by Hugo von Mohl in 1837 as discrete bodies within the green plant cell. In 1883, A. F. W. Schimper would name these bodies as "chloroplastids" (*Chloroplastiden*). In 1884, Eduard Strasburger adopted the term "chloroplasts" (*Chloroplasten*).

Chloroplast Lineages and Evolution

Chloroplasts are one of many types of organelles in the plant cell. They are considered to have originated from cyanobacteria through endosymbiosis—when a eukaryotic cell engulfed a photosynthesizing cyanobacterium that became a permanent resident in the cell. Mitochondria are thought to have come from a similar event, where an aerobic prokaryote was engulfed. This origin of chloroplasts was first suggested by the Russian biologist Konstantin Mereschkowski in 1905 after Andreas Schimper observed in 1883 that chloroplasts closely resemble cyanobacteria. Chloroplasts are only found in plants, algae, and the amoeboid *Paulinella chromatophora*.

Cyanobacterial Ancestor

Cyanobacteria are considered the ancestors of chloroplasts. They are sometimes called blue-green algae even though they are prokaryotes. They are a diverse phylum of bacteria capable of carrying out photosynthesis, and are gram-negative, meaning that they have two cell membranes. Cyanobacteria also contain a peptidoglycan cell wall, which is thicker than in other gram-negative bacteria, and which is located between their two cell membranes. Like chloroplasts, they have thylakoids within. On the thylakoid membranes are photosynthetic pigments, including chlorophyll *a*. Phycobilins are also common cyanobacterial pigments, usually organized into hemispherical phycobilisomes attached to the outside of the thylakoid membranes (phycobilins are not shared with all chloroplasts though).

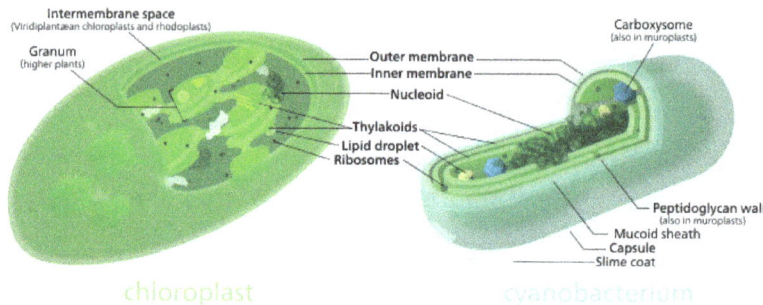

Both chloroplasts and cyanobacteria have a double membrane, DNA, ribosomes, and thylakoids.
Both the chloroplast and cyanobacterium depicted are idealized versions
(the chloroplast is that of a higher plant)—a lot of diversity exists among chloroplasts and cyanobacteria.

Primary Endosymbiosis

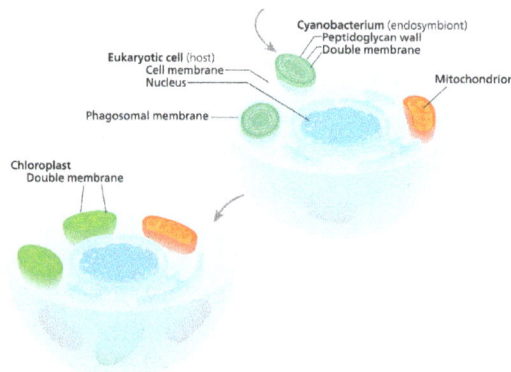

Primary endosymbiosis

A eukaryote with mitochondria engulfed a cyanobacterium in an event of serial primary endosymbiosis, creating a lineage of cells with both organelles. It is important to note that the cyanobacterial endosymbiont already had a double membrane—the phagosomal vacuole-derived membrane was lost.

Somewhere around a billion years ago, a free-living cyanobacterium entered an early eukaryotic cell, either as food or as an internal parasite, but managed to escape the phagocytic vacuole it was contained in. The two innermost lipid-bilayer membranes that surround all chloroplasts correspond to the outer and inner membranes of the ancestral cyanobacterium's gram negative cell wall, and not the phagosomal membrane from the host, which was probably lost. The new cellular resident quickly became an advantage, providing food for the eukaryotic host, which allowed it to live within it. Over time, the cyanobacterium was assimilated, and many of its genes were lost or transferred to the nucleus of the host. From genomes that probably originally contained over 3000 genes only about 130 genes remain in the chloroplasts of contemporary plants. Some of its proteins were then synthesized in the cytoplasm of the host cell, and imported back into the chloroplast (formerly the cyanobacterium). Separately, somewhere around 100 million years ago, it happened again and led to the amoeboid *Paulinella chromatophora*.

This event is called *endosymbiosis*, or "cell living inside another cell". The cell living inside the other cell is called the *endosymbiont*; the endosymbiont is found inside the *host cell*.

Chloroplasts are believed to have arisen after mitochondria, since all eukaryotes contain mitochondria, but not all have chloroplasts. This is called *serial endosymbiosis*—an early eukaryote engulfing the mitochondrion ancestor, and some descendants of it then engulfing the chloroplast ancestor, creating a cell with both chloroplasts and mitochondria.

Whether or not chloroplasts came from a single endosymbiotic event, or many independent engulfments across various eukaryotic lineages, has been long debated. It is now generally held that most organisms with chloroplasts either share a single ancestor or obtained their chloroplast from organisms that share a common ancestor that took in a cyanobacterium 600–1600 million years ago. The exception is the amoeboid *Paulinella chromatophora*, which descends from an ancestor that took in a cyanobacterium 90–140 million years ago.

These chloroplasts, which can be traced back directly to a cyanobacterial ancestor, are known as *primary plastids* ("*plastid*" in this context means almost the same thing as chloroplast). All primary chloroplasts belong to one of four chloroplast lineages—the glaucophyte chloroplast lineage, the amoeboid *Paulinella chromatophora* lineage, the rhodophyte (red algal) chloroplast lineage, or the chloroplastidan (green) chloroplast lineage. The rhodophyte and chloroplastidan lineages are the largest, with chloroplastidan (green) being the one that contains the land plants.

Glaucophyta

The alga *Cyanophora*, a glaucophyte, is thought to be one of the first organisms to contain a chloroplast. The glaucophyte chloroplast group is the smallest of the three primary chloroplast lineages, being found in only 13 species, and is thought to be the one that branched off the earliest. Glaucophytes have chloroplasts that retain a peptidoglycan wall between their double membranes, like their cyanobacterial parent. For this reason, glaucophyte chloroplasts are also known as *muroplasts*. Glaucophyte chloroplasts also contain concentric unstacked thylakoids, which surround a carboxysome – an icosahedral structure that glaucophyte chloroplasts and cyanobacteria keep their carbon fixation enzyme rubisco in. The starch that they synthesize collects outside the chloroplast. Like cyanobacteria, glaucophyte chloroplast thylakoids are studded with light collecting structures called phycobilisomes. For these reasons, glaucophyte chloroplasts are considered a primitive intermediate between cyanobacteria and the more evolved chloroplasts in red algae and plants.

Diversity of red algae Clockwise from top left: *Bornetia secundiflora*, *Peyssonnelia squamaria*, *Cyanidium*, *Laurencia*, *Callophyllis laciniata*. Red algal chloroplasts are characterized by phycobilin pigments which often give them their reddish color.

Rhodophyceae (Red Algae)

The rhodophyte, or red algal chloroplast group is another large and diverse chloroplast lineage. Rhodophyte chloroplasts are also called *rhodoplasts*, literally "red chloroplasts".

Rhodoplasts have a double membrane with an intermembrane space and phycobilin pigments organized into phycobilisomes on the thylakoid membranes, preventing their thylakoids from stacking. Some contain pyrenoids. Rhodoplasts have chlorophyll *a* and phycobilins for photosynthetic pigments; the phycobilin phycoerytherin is responsible for giving many red algae their distinctive red color. However, since they also contain the blue-green chlorophyll *a* and other pigments, many are reddish to purple from the combination. The red phycoerytherin pigment is an adaptation to help red algae catch more sunlight in deep water—as such, some red algae that live in shallow water have less phycoerytherin in their rhodoplasts, and can appear more greenish. Rhodoplasts synthesize a form of starch called floridean starch, which collects into granules outside the rhodoplast, in the cytoplasm of the red alga.

Chloroplastida (Green Algae and Plants)

The chloroplastidan chloroplasts, or green chloroplasts, are another large, highly diverse primary chloroplast lineage. Their host organisms are commonly known as the green algae and land plants.

They differ from glaucophyte and red algal chloroplasts in that they have lost their phycobilisomes, and contain chlorophyll *b* instead. Most green chloroplasts are (obviously) green, though some aren't, like some forms of *Hæmatococcus pluvialis*, due to accessory pigments that override the chlorophylls' green colors. Chloroplastidan chloroplasts have lost the peptidoglycan wall between their double membrane, leaving an intermembrane space. Some plants seem to have kept the genes for the synthesis of the peptidoglycan layer, though they've been repurposed for use in chloroplast division instead.

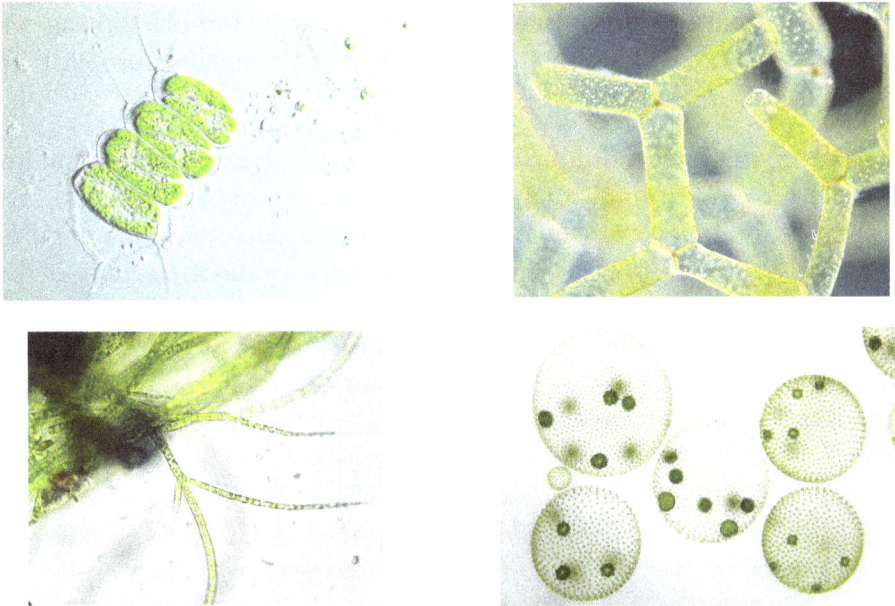

Diversity of green algae Clockwise from top left: *Scenedesmus, Micrasterias, Hydrodictyon, Volvox, Stigeoclonium*. Green algal chloroplasts are characterized by their pigments chlorophyll *a* and chlorophyll *b* which give them their green color.

Most of the chloroplasts depicted here are green chloroplasts.

Green algae and plants keep their starch *inside* their chloroplasts, and in plants and some algae, the chloroplast thylakoids are arranged in grana stacks. Some green algal chloroplasts contain a structure called a pyrenoid, which is functionally similar to the glaucophyte carboxysome in that it is where rubisco and CO_2 are concentrated in the chloroplast.

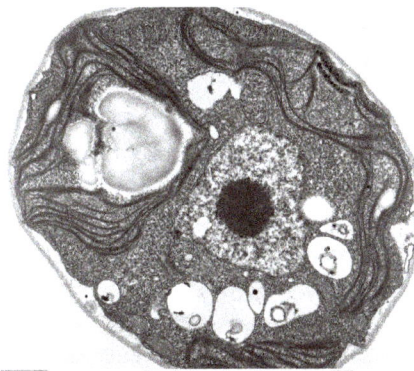

Transmission electron micrograph of *Chlamydomonas reinhardtii*, a green alga that contains a pyrenoid surrounded by starch.

Helicosporidium

Helicosporidium is a genus of nonphotosynthetic parasitic green algae that is thought to contain a vestigial chloroplast. Genes from a chloroplast and nuclear genes indicating the presence of a chloroplast have been found in Helicosporidium even if nobody's seen the chloroplast itself.

Secondary and Tertiary Endosymbiosis

Many other organisms obtained chloroplasts from the primary chloroplast lineages through secondary endosymbiosis—engulfing a red or green alga that contained a chloroplast. These chloroplasts are known as secondary plastids.

While primary chloroplasts have a double membrane from their cyanobacterial ancestor, secondary chloroplasts have additional membranes outside of the original two, as a result of the secondary endosymbiotic event, when a nonphotosynthetic eukaryote engulfed a chloroplast-containing alga but failed to digest it—much like the cyanobacterium at the beginning of this story. The engulfed alga was broken down, leaving only its chloroplast, and sometimes its cell membrane and nucleus, forming a chloroplast with three or four membranes—the two cyanobacterial membranes, sometimes the eaten alga's cell membrane, and the phagosomal vacuole from the host's cell membrane.

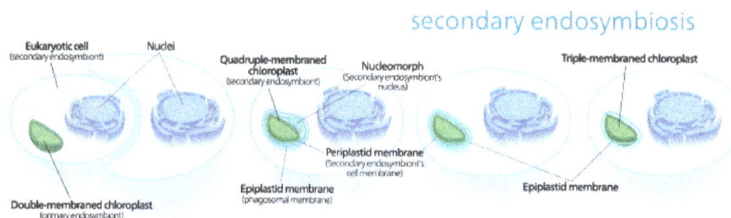

Secondary endosymbiosis consisted of a eukaryotic alga being engulfed by another eukaryote, forming a chloroplast with three or four membranes.

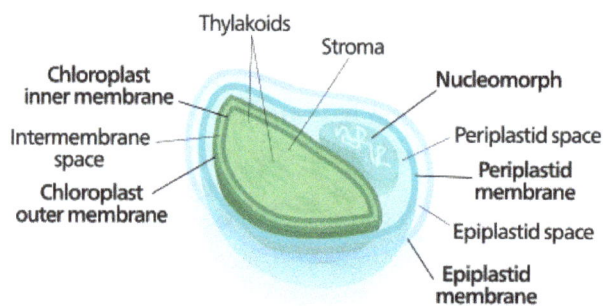

Diagram of a four membraned chloroplast containing a nucleomorph.

The genes in the phagocytosed eukaryote's nucleus are often transferred to the secondary host's nucleus. Cryptomonads and chlorarachniophytes retain the phagocytosed eukaryote's nucleus, an object called a nucleomorph, located between the second and third membranes of the chloroplast.

All secondary chloroplasts come from green and red algae—no secondary chloroplasts from glaucophytes have been observed, probably because glaucophytes are relatively rare in nature, making them less likely to have been taken up by another eukaryote.

Green Algal Derived Chloroplasts

Green algae have been taken up by the euglenids, chlorarachniophytes, a lineage of dinoflagellates, and possibly the ancestor of the chromalveolates in three or four separate engulfments. Many green algal derived chloroplasts contain pyrenoids, but unlike chloroplasts in their green algal ancestors, starch collects in granules outside the chloroplast.

Euglena, a euglenophyte, contains secondary chloroplasts from green algae.

Euglenophytes

Euglenophytes are a group of common flagellated protists that contain chloroplasts derived from a green alga. Euglenophyte chloroplasts have three membranes—it is thought that the membrane of the primary endosymbiont was lost, leaving the cyanobacterial membranes, and the secondary host's phagosomal membrane. Euglenophyte chloroplasts have a pyrenoid and thylakoids stacked in groups of three. Starch is stored in the form of paramylon, which is contained in membrane-bound granules in the cytoplasm of the euglenophyte.

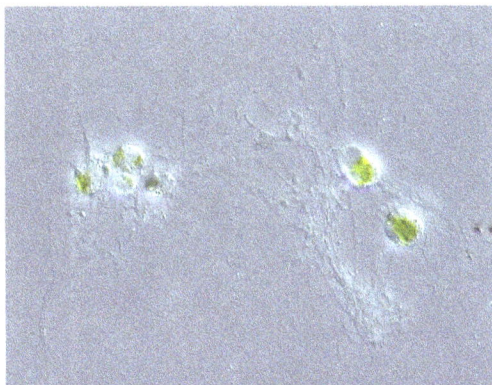

Chlorarachnion reptans is a chlorarachniophyte. Chlorarachniophytes replaced their original red algal endosymbiont with a green alga.

Chlorarachniophytes

Chlorarachniophytes are a rare group of organisms that also contain chloroplasts derived from green algae, though their story is more complicated than that of the euglenophytes. The ancestor of chlorarachniophytes is thought to have been a chromalveolate, a eukaryote with a *red* algal derived chloroplast. It is then thought to have lost its first red algal chloroplast, and later engulfed a green alga, giving it its second, green algal derived chloroplast.

Chlorarachniophyte chloroplasts are bounded by four membranes, except near the cell membrane, where the chloroplast membranes fuse into a double membrane. Their thylakoids are arranged in loose stacks of three. Chlorarachniophytes have a form of starch called chrysolaminarin, which they store in the cytoplasm, often collected around the chloroplast pyrenoid, which bulges into the cytoplasm.

Chlorarachniophyte chloroplasts are notable because the green alga they are derived from has not been completely broken down—its nucleus still persists as a nucleomorph found between the second and third chloroplast membranes—the periplastid space, which corresponds to the green alga's cytoplasm.

Early Chromalveolates

Recent research has suggested that the ancestor of the chromalveolates acquired a green algal prasinophyte endosymbiont. The green algal derived chloroplast was lost and replaced with a red algal derived chloroplast, but not before contributing some of its genes to the early chromalveolate's nucleus. The presence of both green algal and red algal genes in chromalveolates probably helps them thrive under fluctuating light conditions.

Red Algal Derived Chloroplasts (Chromalveolate Chloroplasts)

Like green algae, red algae have also been taken up in secondary endosymbiosis, though it is thought that all red algal derived chloroplasts are descended from a single red alga that was engulfed by an early chromalveolate, giving rise to the chromalveolates, some of which, like the ciliates, subsequently lost the chloroplast. This is still debated though.

Pyrenoids and stacked thylakoids are common in chromalveolate chloroplasts, and the outermost membrane of many are continuous with the rough endoplasmic reticulum and studded with ribosomes. They have lost their phycobilisomes and exchanged them for chlorophyll *c*, which isn't found in primary red algal chloroplasts themselves.

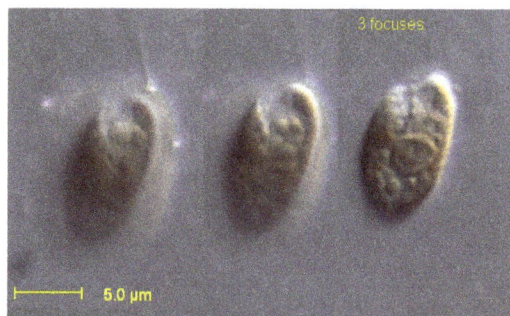

Rhodomonas salina is a cryptophyte.

Cryptophytes

Cryptophytes, or cryptomonads are a group of algae that contain a red-algal derived chloroplast. Cryptophyte chloroplasts contain a nucleomorph that superficially resembles that of the chlorarachniophytes. Cryptophyte chloroplasts have four membranes, the outermost of which is continuous with the rough endoplasmic reticulum. They synthesize ordinary starch, which is stored in granules found in the periplastid space—outside the original double membrane, in the place that corresponds to the red alga's cytoplasm. Inside cryptophyte chloroplasts is a pyrenoid and thylakoids in stacks of two.

Their chloroplasts do not have phycobilisomes, but they do have phycobilin pigments which they keep in their thylakoid space, rather than anchored on the outside of their thylakoid membranes.

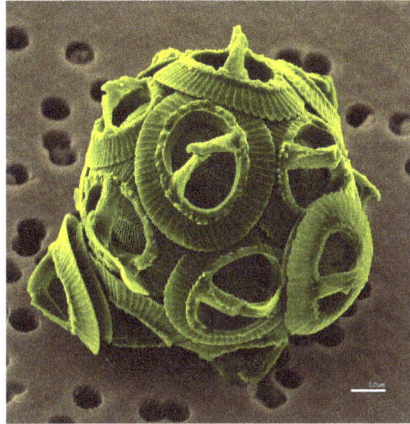

Scanning electron micrograph of *Gephyrocapsa oceanica*, a haptophyte.

Haptophytes

Haptophytes are similar and closely related to cryptophytes, and are thought to be the first chromalveolates to branch off. Their chloroplasts lack a nucleomorph, their thylakoids are in stacks of three, and they synthesize chrysolaminarin sugar, which they store completely outside of the chloroplast, in the cytoplasm of the haptophyte.

Heterokontophytes (Stramenopiles)

The photosynthetic pigments present in their chloroplasts give diatoms a greenish-brown color.

The heterokontophytes, also known as the stramenopiles, are a very large and diverse group of algae that also contain red algal derived chloroplasts. Heterokonts include the diatoms and the brown algae, golden algae, and yellow-green algae.

Heterokont chloroplasts are very similar to haptophyte chloroplasts, containing a pyrenoid, triplet thylakoids, and with some exceptions, having an epiplastid membrane connected to the endoplasmic reticulum. Like haptophytes, heterokontophytes store sugar in chrysolaminarin granules in the cytoplasm. Heterokontophyte chloroplasts contain chlorophyll a and with a few exceptions chlorophyll c, but also have carotenoids which give them their many colors.

Apicomplexans

Apicomplexans are another group of chromalveolates. Like the helicosproidia, they're parasitic, and have a nonphotosynthetic chloroplast. They were once thought to be related to the helicosproidia, but it is now known that the helicosproida are green algae rather than chromalveolates. The apicomplexans include *Plasmodium*, the malaria parasite. Many apicomplexans keep a vestigial red algal derived chloroplast called an apicoplast, which they inherited from their ancestors. Other apicomplexans like *Cryptosporidium* have lost the chloroplast completely. Apicomplexans store their energy in amylopectin starch granules that are located in their cytoplasm, even though they are nonphotosynthetic.

Apicoplasts have lost all photosynthetic function, and contain no photosynthetic pigments or true thylakoids. They are bounded by four membranes, but the membranes are not connected to the endoplasmic reticulum. The fact that apicomplexans still keep their nonphotosynthetic chloroplast around demonstrates how the chloroplast carries out important functions other than photosynthesis. Plant chloroplasts provide plant cells with many important things besides sugar, and apicoplasts are no different—they synthesize fatty acids, isopentenyl pyrophosphate, iron-sulfur clusters, and carry out part of the heme pathway. This makes the apicoplast an attractive target for drugs to cure apicomplexan-related diseases. The most important apicoplast function is isopentenyl pyrophosphate synthesis—in fact, apicomplexans die when something interferes with this apicoplast function, and when apicomplexans are grown in an isopentenyl pyrophosphate-rich medium, they dump the organelle.

Dinophytes

The dinoflagellates are yet another very large and diverse group of protists, around half of which are (at least partially) photosynthetic.

Most dinophyte chloroplasts are secondary red algal derived chloroplasts, like other chromalveolate chloroplasts. Many other dinophytes have lost the chloroplast (becoming the nonphotosynthetic kind of dinoflagellate), or replaced it though *tertiary* endosymbiosis—the engulfment of another chromalveolate containing a red algal derived chloroplast. Others replaced their original chloroplast with a green algal derived one.

Most dinophyte chloroplasts contain at least the photosynthetic pigments chlorophyll a, chlorophyll c_2, *beta*-carotene, and at least one dinophyte-unique xanthophyll (peridinin, dinoxanthin, or diadinoxanthin), giving many a golden-brown color. All dinophytes store starch in their cytoplasm, and most have chloroplasts with thylakoids arranged in stacks of three.

Peridinin-containing Dinophyte Chloroplast

The most common dinophyte chloroplast is the peridinin-type chloroplast, characterized by the carotenoid pigment peridinin in their chloroplasts, along with chlorophyll a and chlorophyll c_2.

Peridinin is not found in any other group of chloroplasts. The peridinin chloroplast is bounded by three membranes (occasionally two), having lost the red algal endosymbiont's original cell membrane. The outermost membrane is not connected to the endoplasmic reticulum. They contain a pyrenoid, and have triplet-stacked thylakoids. Starch is found outside the chloroplast. An important feature of these chloroplasts is that their chloroplast DNA is highly reduced and fragmented into many small circles. Most of the genome has migrated to the nucleus, and only critical photosynthesis-related genes remain in the chloroplast.

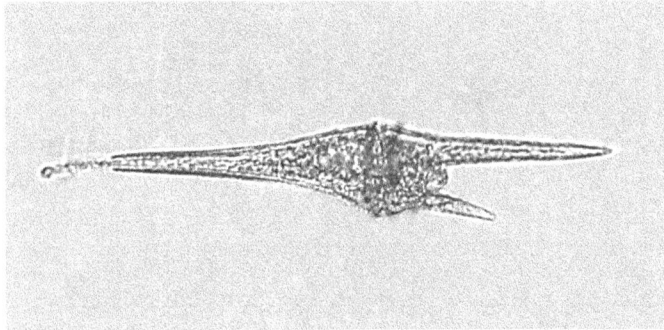

Ceratium furca, a peridinin-containing dinophyte.

The peridinin chloroplast is thought to be the dinophytes' "original" chloroplast, which has been lost, reduced, replaced, or has company in several other dinophyte lineages.

Fucoxanthin-containing Dinophyte Chloroplasts (Haptophyte Endosymbizonts)

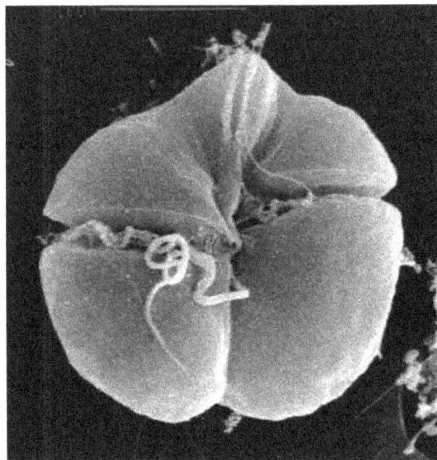

Karenia brevis is a fucoxanthin-containing dynophyte responsible for algal blooms called "red tides".

The fucoxanthin dinophyte lineages (including *Karlodinium* and *Karenia*) lost their original red algal derived chloroplast, and replaced it with a new chloroplast derived from a haptophyte endosymbiont. *Karlodinium* and *Karenia* probably took up different heterokontophytes. Because the haptophyte chloroplast has four membranes, tertiary endosymbiosis would be expected to create a six membraned chloroplast, adding the haptophyte's cell membrane and the dinophyte's phagosomal vacuole. However, the haptophyte was heavily reduced, stripped of a few membranes and its nucleus, leaving only its chloroplast (with its original double membrane), and possibly one or two additional membranes around it.

Fucoxanthin-containing chloroplasts are characterized by having the pigment fucoxanthin (actually 19'-hexanoyloxy-fucoxanthin and/or 19'-butanoyloxy-fucoxanthin) and no peridinin. Fucoxanthin is also found in haptophyte chloroplasts, providing evidence of ancestry.

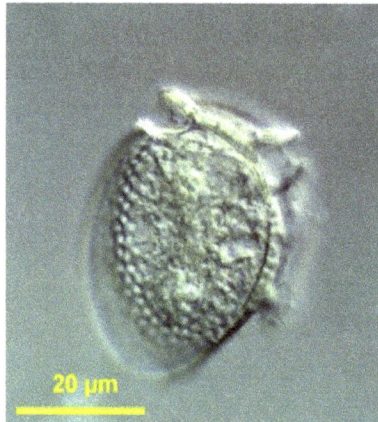

Dinophysis acuminata has chloroplasts taken from a cryptophyte.

Cryptophyte Derived Dinophyte Chloroplast

Members of the genus *Dinophysis* have a phycobilin-containing chloroplast taken from a cryptophyte. However, the cryptophyte is not an endosymbiont—only the chloroplast seems to have been taken, and the chloroplast has been stripped of its nucleomorph and outermost two membranes, leaving just a two-membraned chloroplast. Cryptophyte chloroplasts require their nucleomorph to maintain themselves, and *Dinophysis* species grown in cell culture alone cannot survive, so it is possible (but not confirmed) that the *Dinophysis* chloroplast is a kleptoplast—if so, *Dinophysis* chloroplasts wear out and *Dinophysis* species must continually engulf cryptophytes to obtain new chloroplasts to replace the old ones.

Diatom Derived Dinophyte Chloroplasts

Some dinophytes, like *Kryptoperidinium* and *Durinskia* have a diatom (heterokontophyte) derived chloroplast. These chloroplasts are bounded by up to *five* membranes, (depending on whether you count the entire diatom endosymbiont as the chloroplast, or just the red algal derived chloroplast inside it). The diatom endosymbiont has been reduced relatively little—it still retains its original mitochondria, and has endoplasmic reticulum, ribosomes, a nucleus, and of course, red algal derived chloroplasts—practically a complete cell, all inside the host's endoplasmic reticulum lumen. However the diatom endosymbiont can't store its own food—its starch is found in granules in the dinophyte host's cytoplasm instead. The diatom endosymbiont's nucleus is present, but it probably can't be called a nucleomorph because it shows no sign of genome reduction, and might have even been *expanded*. Diatoms have been engulfed by dinoflagellates at least three times.

The diatom endosymbiont is bounded by a single membrane, inside it are chloroplasts with four membranes. Like the diatom endosymbiont's diatom ancestor, the chloroplasts have triplet thylakoids and pyrenoids.

In some of these genera, the diatom endosymbiont's chloroplasts aren't the only chloroplasts in the dinophyte. The original three-membraned peridinin chloroplast is still around, converted to an eyespot.

Prasinophyte (Green Algal) Derived Dinophyte Chloroplast

Lepidodinium viride and its close relatives are dinophytes that lost their original peridinin chloroplast and replaced it with a green algal derived chloroplast (more specifically, a prasinophyte). *Lepidodinium* is the only dinophyte that has a chloroplast that's not from the rhodoplast lineage. The chloroplast is surrounded by two membranes and has no nucleomorph—all the nucleomorph genes have been transferred to the dinophyte nucleus. The endosymbiotic event that led to this chloroplast was serial secondary endosymbiosis rather than tertiary endosymbiosis—the endosymbiont was a green alga containing a primary chloroplast (making a secondary chloroplast).

Chromatophores

While most chloroplasts originate from that first set of endosymbiotic events, *Paulinella chromatophora* is an exception that acquired a photosynthetic cyanobacterial endosymbiont more recently. It is not clear whether that symbiont is closely related to the ancestral chloroplast of other eukaryotes. Being in the early stages of endosymbiosis, *Paulinella chromatophora* can offer some insights into how chloroplasts evolved. *Paulinella* cells contain one or two sausage shaped blue-green photosynthesizing structures called chromatophores, descended from the cyanobacterium *Synechococcus*. Chromatophores cannot survive outside their host. Chromatophore DNA is about a million base pairs long, containing around 850 protein encoding genes—far less than the three million base pair *Synechococcus* genome, but much larger than the approximately 150,000 base pair genome of the more assimilated chloroplast. Chromatophores have transferred much less of their DNA to the nucleus of their host. About 0.3–0.8% of the nuclear DNA in *Paulinella* is from the chromatophore, compared with 11–14% from the chloroplast in plants.

Kleptoplastidy

In some groups of mixotrophic protists, like some dinoflagellates, chloroplasts are separated from a captured alga or diatom and used temporarily. These klepto chloroplasts may only have a lifetime of a few days and are then replaced.

Chloroplast DNA

Chloroplasts have their own DNA, often abbreviated as ctDNA, or cpDNA. It is also known as the plastome. Its existence was first proved in 1962, and first sequenced in 1986—when two Japanese research teams sequenced the chloroplast DNA of liverwort and tobacco. Since then, hundreds of chloroplast DNAs from various species have been sequenced, but they're mostly those of land plants and green algae—glaucophytes, red algae, and other algal groups are extremely underrepresented, potentially introducing some bias in views of "typical" chloroplast DNA structure and content.

Molecular Structure

Chloroplast DNA Interactive gene map of chloroplast DNA from *Nicotiana tabacum*. Segments with labels on the inside reside on the B strand of DNA, segments with labels on the outside are on the A strand. Notches indicate introns.

With few exceptions, most chloroplasts have their entire chloroplast genome combined into a single large circular DNA molecule, typically 120,000–170,000 base pairs long. They can have a contour length of around 30–60 micrometers, and have a mass of about 80–130 million daltons.

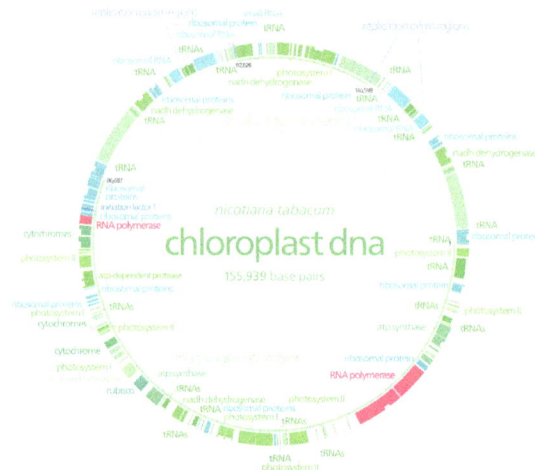

While usually thought of as a circular molecule, there is some evidence that chloroplast DNA molecules more often take on a linear shape.

Inverted Repeats

Many chloroplast DNAs contain two *inverted repeats*, which separate a long single copy section (LSC) from a short single copy section (SSC). While a given pair of inverted repeats are rarely completely identical, they are always very similar to each other, apparently resulting from concerted evolution.

The inverted repeats vary wildly in length, ranging from 4,000 to 25,000 base pairs long each and containing as few as four or as many as over 150 genes. Inverted repeats in plants tend to be at the upper end of this range, each being 20,000–25,000 base pairs long.

The inverted repeat regions are highly conserved among land plants, and accumulate few mutations. Similar inverted repeats exist in the genomes of cyanobacteria and the other two chloroplast lineages (glaucophyta and rhodophyceae), suggesting that they predate the chloroplast, though some chloroplast DNAs have since lost or flipped the inverted repeats (making them direct repeats). It is possible that the inverted repeats help stabilize the rest of the chloroplast genome, as chloroplast DNAs which have lost some of the inverted repeat segments tend to get rearranged more.

Nucleoids

New chloroplasts may contain up to 100 copies of their DNA, though the number of chloroplast DNA copies decreases to about 15–20 as the chloroplasts age. They are usually packed into nucleoids, which can contain several identical chloroplast DNA rings. Many nucleoids can be found in each chloroplast. In primitive red algae, the chloroplast DNA nucleoids are clustered in the center of the chloroplast, while in green plants and green algae, the nucleoids are dispersed throughout the stroma.

Though chloroplast DNA is not associated with true histones, in red algae, similar proteins that tightly pack each chloroplast DNA ring into a nucleoid have been found.

DNA Replication

The Leading Model of CpDNA Replication

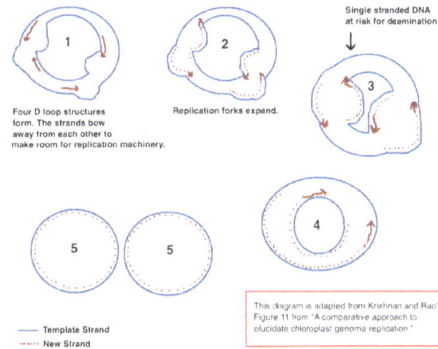

Chloroplast DNA replication via multiple D loop mechanisms. Adapted from
Krishnan NM, Rao BJ's paper "A comparative approach to elucidate chloroplast genome replication."

The mechanism for chloroplast DNA (cpDNA) replication has not been conclusively determined, but two main models have been proposed. Scientists have attempted to observe chloroplast replication via electron microscopy since the 1970s. The results of the microscopy experiments led to the idea that chloroplast DNA replicates using a double displacement loop (D-loop). As the D-loop moves through the circular DNA, it adopts a theta intermediary form, also known as a Cairns replication intermediate, and completes replication with a rolling circle mechanism. Transcription starts at specific points of origin. Multiple replication forks open up, allowing replication machinery to transcribe the DNA. As replication continues, the forks grow and eventually converge. The new cpDNA structures separate, creating daughter cpDNA chromosomes.

In addition to the early microscopy experiments, this model is also supported by the amounts of deamination seen in cpDNA. Deamination occurs when an amino group is lost and is a mutation that often results in base changes. When adenine is deaminated, it becomes hypoxanthine. Hypoxanthine can bind to cytosine, and when the XC base pair is replicated, it becomes a GC (thus, an A → G base change).

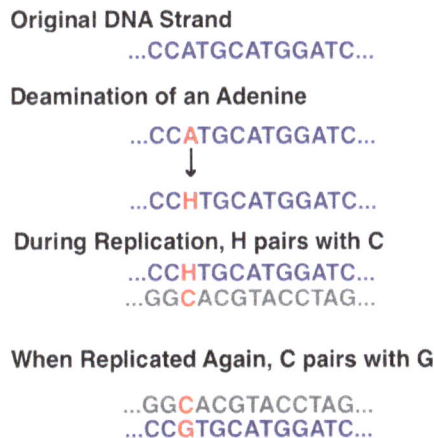

Original DNA Strand
...CCATGCATGGATC...

Deamination of an Adenine
...CCATGCATGGATC...
↓
...CCHTGCATGGATC...

During Replication, H pairs with C
...CCHTGCATGGATC...
...GGCACGTACCTAG...

When Replicated Again, C pairs with G
...GGCACGTACCTAG...
...CCGTGCATGGATC...

Over time, base changes in the DNA sequence can arise from deamination mutations.
When adenine is deaminated, it becomes hypoxanthine, which can pair with cytosine. During replication,
the cytosine will pair with guanine, causing an A --> G base change.

Deamination

In cpDNA, there are several A → G deamination gradients. DNA becomes susceptible to deamination events when it is single stranded. When replication forks form, the strand not being copied is single stranded, and thus at risk for A → G deamination. Therefore, gradients in deamination indicate that replication forks were most likely present and the direction that they initially opened (the highest gradient is most likely nearest the start site because it was single stranded for the longest amount of time). This mechanism is still the leading theory today; however, a second theory suggests that most cpDNA is actually linear and replicates through homologous recombination. It further contends that only a minority of the genetic material is kept in circular chromosomes while the rest is in branched, linear, or other complex structures.

Alternative Model of Replication

One of competing model for cpDNA replication asserts that most cpDNA is linear and participates in homologous recombination and replication structures similar to bacteriophage T4. It has been established that some plants have linear cpDNA, such as maize, and that more species still contain complex structures that scientists do not yet understand. When the original experiments on cpDNA were performed, scientists did notice linear structures; however, they attributed these linear forms to broken circles. If the branched and complex structures seen in cpDNA experiments are real and not artifacts of concatenated circular DNA or broken circles, then a D-loop mechanism of replication is insufficient to explain how those structures would replicate. At the same time, homologous recombination does not expand the multiple A --> G gradients seen in plastomes. Because of the failure to explain the deamination gradient as well as the numerous plant species that have been shown to have circular cpDNA, the predominant theory continues to hold that most cpDNA is circular and most likely replicates via a D loop mechanism.

Gene Content and Protein Synthesis

The chloroplast genome most commonly includes around 100 genes that code for a variety of things, mostly to do with the protein pipeline and photosynthesis. As in prokaryotes, genes in chloroplast DNA are organized into operons. Interestingly though, unlike prokaryotic DNA molecules, chloroplast DNA molecules contain introns (plant mitochondrial DNAs do too, but not human mtDNAs).

Among land plants, the contents of the chloroplast genome are fairly similar.

Chloroplast Genome Reduction and Gene Transfer

Over time, many parts of the chloroplast genome were transferred to the nuclear genome of the host, a process called *endosymbiotic gene transfer*. As a result, the chloroplast genome is heavily reduced compared to that of free-living cyanobacteria. Chloroplasts may contain 60–100 genes whereas cyanobacteria often have more than 1500 genes in their genome. Recently, a plastid without a genome was found, demonstrating chloroplasts can lose their genome during endosymbiotic the gene transfer process.

Endosymbiotic gene transfer is how we know about the lost chloroplasts in many chromalveolate lineages. Even if a chloroplast is eventually lost, the genes it donated to the former host's nucleus persist, providing evidence for the lost chloroplast's existence. For example, while diatoms (a heterokontophyte) now have a red algal derived chloroplast, the presence of many green algal genes in the diatom nucleus provide evidence that the diatom ancestor (probably the ancestor of all chromalveolates too) had a green algal derived chloroplast at some point, which was subsequently replaced by the red chloroplast.

In land plants, some 11–14% of the DNA in their nuclei can be traced back to the chloroplast, up to 18% in *Arabidopsis*, corresponding to about 4,500 protein-coding genes. There have been a few recent transfers of genes from the chloroplast DNA to the nuclear genome in land plants.

Of the approximately 3000 proteins found in chloroplasts, some 95% of them are encoded by nuclear genes. Many of the chloroplast's protein complexes consist of subunits from both the chloroplast genome and the host's nuclear genome. As a result, protein synthesis must be coordinated between the chloroplast and the nucleus. The chloroplast is mostly under nuclear control, though chloroplasts can also give out signals regulating gene expression in the nucleus, called *retrograde signaling*.

Protein Synthesis

Protein synthesis within chloroplasts relies on two RNA polymerases. One is coded by the chloroplast DNA, the other is of nuclear origin. The two RNA polymerases may recognize and bind to different kinds of promoters within the chloroplast genome. The ribosomes in chloroplasts are similar to bacterial ribosomes.

Protein Targeting and Import

Because so many chloroplast genes have been moved to the nucleus, many proteins that would originally have been translated in the chloroplast are now synthesized in the cytoplasm of the plant cell. These proteins must be directed back to the chloroplast, and imported through at least two chloroplast membranes.

Curiously, around half of the protein products of transferred genes aren't even targeted back to the chloroplast. Many became exaptations, taking on new functions like participating in cell division, protein routing, and even disease resistance. A few chloroplast genes found new homes in the mitochondrial genome—most became nonfunctional pseudogenes, though a few tRNA genes still work in the mitochondrion. Some transferred chloroplast DNA protein products get directed to the secretory pathway though it should be noted that many secondary plastids are bounded by an outermost membrane derived from the host's cell membrane, and therefore topologically outside of the cell, because to reach the chloroplast from the cytosol, you have to cross the cell membrane, just like if you were headed for the extracellular space. In those cases, chloroplast-targeted proteins do initially travel along the secretory pathway.

Because the cell acquiring a chloroplast already had mitochondria (and peroxisomes, and a cell membrane for secretion), the new chloroplast host had to develop a unique protein targeting system to avoid having chloroplast proteins being sent to the wrong organelle.

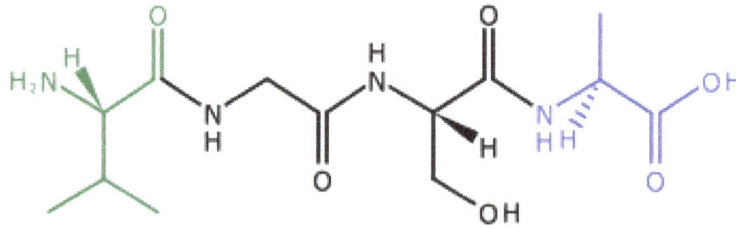

The two ends of a polypeptide are called the N-terminus, or *amino end*, and the C-terminus, or *carboxyl end*. This polypeptide has four amino acids linked together. At the left is the N-terminus, with its amino (H_2N) group in green. The blue C-terminus, with its carboxyl group (CO_2H) is at the right.

In most, but not all cases, nuclear-encoded chloroplast proteins are translated with a *cleavable transit peptide* that's added to the N-terminus of the protein precursor. Sometimes the transit sequence is found on the C-terminus of the protein, or within the functional part of the protein.

Transport Proteins and Membrane Translocons

After a chloroplast polypeptide is synthesized on a ribosome in the cytosol, an enzyme specific to chloroplast proteins phosphorylates, or adds a phosphate group to many (but not all) of them in their transit sequences. Phosphorylation helps many proteins bind the polypeptide, keeping it from folding prematurely. This is important because it prevents chloroplast proteins from assuming their active form and carrying out their chloroplast functions in the wrong place—the cytosol. At the same time, they have to keep just enough shape so that they can be recognized by the chloroplast. These proteins also help the polypeptide get imported into the chloroplast.

From here, chloroplast proteins bound for the stroma must pass through two protein complexes—the TOC complex, or *translocon on the outer chloroplast membrane*, and the TIC translocon, or *translocon on the inner chloroplast membrane translocon*. Chloroplast polypeptide chains probably often travel through the two complexes at the same time, but the TIC complex can also retrieve preproteins lost in the intermembrane space.

Structure

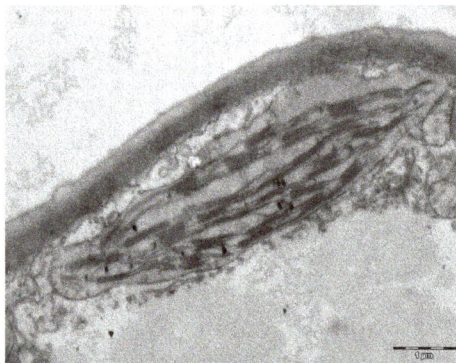

Transmission electron microscope image of a chloroplast.
Grana of thylakoids and their connecting lamellae are clearly visible.

In land plants, chloroplasts are generally lens-shaped, 3–10 µm in diameter and 1–3 µm thick. Corn seedling chloroplasts are ≈20 µm³ in volume. Greater diversity in chloroplast shapes exists among the algae, which often contain a single chloroplast that can be shaped like a net (e.g., *Oe-*

dogonium), a cup (e.g., *Chlamydomonas*), a ribbon-like spiral around the edges of the cell (e.g., *Spirogyra*), or slightly twisted bands at the cell edges (e.g., *Sirogonium*). Some algae have two chloroplasts in each cell; they are star-shaped in *Zygnema*, or may follow the shape of half the cell in order Desmidiales. In some algae, the chloroplast takes up most of the cell, with pockets for the nucleus and other organelles, for example, some species of *Chlorella* have a cup-shaped chloroplast that occupies much of the cell.

All chloroplasts have at least three membrane systems—the outer chloroplast membrane, the inner chloroplast membrane, and the thylakoid system. Chloroplasts that are the product of secondary endosymbiosis may have additional membranes surrounding these three. Inside the outer and inner chloroplast membranes is the chloroplast stroma, a semi-gel-like fluid that makes up much of a chloroplast's volume, and in which the thylakoid system floats.

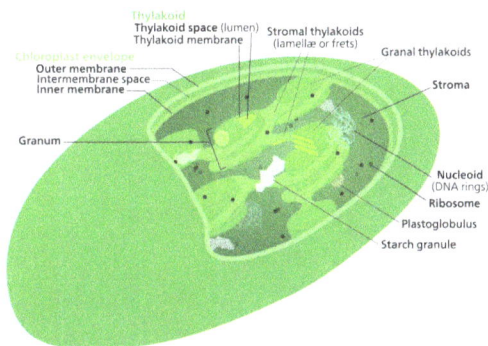

Chloroplast ultrastructure *(interactive diagram)* Chloroplasts have at least three distinct membrane systems, and a variety of things can be found in their stroma.

There are some common misconceptions about the outer and inner chloroplast membranes. The fact that chloroplasts are surrounded by a double membrane is often cited as evidence that they are the descendants of endosymbiotic cyanobacteria. This is often interpreted as meaning the outer chloroplast membrane is the product of the host's cell membrane infolding to form a vesicle to surround the ancestral cyanobacterium—which is not true—both chloroplast membranes are homologous to the cyanobacterium's original double membranes.

The chloroplast double membrane is also often compared to the mitochondrial double membrane. This is not a valid comparison—the inner mitochondria membrane is used to run proton pumps and carry out oxidative phosphorylation across to generate ATP energy. The only chloroplast structure that can considered analogous to it is the internal thylakoid system. Even so, in terms of "in-out", the direction of chloroplast H^+ ion flow is in the opposite direction compared to oxidative phosphorylation in mitochondria. In addition, in terms of function, the inner chloroplast membrane, which regulates metabolite passage and synthesizes some materials, has no counterpart in the mitochondrion.

Outer Chloroplast Membrane

The outer chloroplast membrane is a semi-porous membrane that small molecules and ions can easily diffuse across. However, it is not permeable to larger proteins, so chloroplast polypeptides being synthesized in the cell cytoplasm must be transported across the outer chloroplast membrane by the TOC complex, or translocon on the outer chloro*last* membrane.

The chloroplast membranes sometimes protrude out into the cytoplasm, forming a stromule, or stroma-containing tubule. Stromules are very rare in chloroplasts, and are much more common in other plastids like chromoplasts and amyloplasts in petals and roots, respectively. They may exist to increase the chloroplast's surface area for cross-membrane transport, because they are often branched and tangled with the endoplasmic reticulum. When they were first observed in 1962, some plant biologists dismissed the structures as artifactual, claiming that stromules were just oddly shaped chloroplasts with constricted regions or dividing chloroplasts. However, there is a growing body of evidence that stromules are functional, integral features of plant cell plastids, not merely artifacts.

Intermembrane Space and Peptidoglycan Wall

Instead of an intermembrane space, glaucophyte algae have a
peptidoglycan wall between their inner and outer chloroplast membranes.

Usually, a thin intermembrane space about 10–20 nanometers thick exists between the outer and inner chloroplast membranes.

Glaucophyte algal chloroplasts have a peptidoglycan layer between the chloroplast membranes. It corresponds to the peptidoglycan cell wall of their cyanobacterial ancestors, which is located between their two cell membranes. These chloroplasts are called *muroplasts* (from Latin *"mura"*, meaning "wall"). Other chloroplasts have lost the cyanobacterial wall, leaving an intermembrane space between the two chloroplast envelope membranes.

Inner Chloroplast Membrane

The inner chloroplast membrane borders the stroma and regulates passage of materials in and out of the chloroplast. After passing through the TOC complex in the outer chloroplast membrane, polypeptides must pass through the TIC complex *(translocon on the inner chloroplast membrane)* which is located in the inner chloroplast membrane.

In addition to regulating the passage of materials, the inner chloroplast membrane is where fatty acids, lipids, and carotenoids are synthesized.

Peripheral Reticulum

Some chloroplasts contain a structure called the chloroplast peripheral reticulum. It is often found in the chloroplasts of C_4 plants, though it has also been found in some C_3 angiosperms, and even some gymnosperms. The chloroplast peripheral reticulum consists of a maze of membranous tubes and vesicles continuous with the inner chloroplast membrane that extends into the internal stromal fluid of the chloroplast. Its purpose is thought to be to increase the chloroplast's surface area for cross-membrane transport between its stroma and the cell cytoplasm. The small vesicles sometimes observed may serve as transport vesicles to shuttle stuff between the thylakoids and intermembrane space.

Stroma

The protein-rich, alkaline, aqueous fluid within the inner chloroplast membrane and outside of the thylakoid space is called the stroma, which corresponds to the cytosol of the original cyanobacterium. Nucleoids of chloroplast DNA, chloroplast ribosomes, the thylakoid system with plastoglobuli, starch granules, and many proteins can be found floating around in it. The Calvin cycle, which fixes CO_2 into sugar takes place in the stroma.

Chloroplast Ribosomes

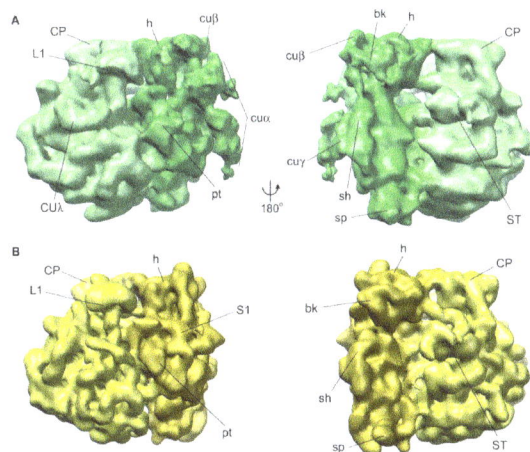

Chloroplast ribosomes Comparison of a chloroplast ribosome (green) and a bacterial ribosome (yellow). Important features common to both ribosomes and chloroplast-unique features are labeled.

Chloroplasts have their own ribosomes, which they use to synthesize a small fraction of their proteins. Chloroplast ribosomes are about two-thirds the size of cytoplasmic ribosomes (around 17 nm vs 25 nm). They take mRNAs transcribed from the chloroplast DNA and translate them into protein. While similar to bacterial ribosomes, chloroplast translation is more complex than in bacteria, so chloroplast ribosomes include some chloroplast-unique features. Small subunit ribosomal RNAs in several Chlorophyta and euglenid chloroplasts lack motifs for shine-dalgarno sequence recognition, which is considered essential for translation initiation in most chloroplasts and prokaryotes. Such loss is also rarely observed in other plastids and prokaryotes.

Plastoglobuli

Plastoglobuli (singular *plastoglobulus*, sometimes spelled *plastoglobule(s)*), are spherical bubbles of lipids and proteins about 45–60 nanometers across. They are surrounded by a lipid monolayer. Plastoglobuli are found in all chloroplasts, but become more common when the chloroplast is under oxidative stress, or when it ages and transitions into a gerontoplast. Plastoglobuli also exhibit a greater size variation under these conditions. They are also common in etioplasts, but decrease in number as the etioplasts mature into chloroplasts.

Plastoglubuli contain both structural proteins and enzymes involved in lipid synthesis and metabolism. They contain many types of lipids including plastoquinone, vitamin E, carotenoids and chlorophylls.

Plastoglobuli were once thought to be free-floating in the stroma, but it is now thought that they are permanently attached either to a thylakoid or to another plastoglobulus attached to a thylakoid, a configuration that allows a plastoglobulus to exchange its contents with the thylakoid network. In normal green chloroplasts, the vast majority of plastoglobuli occur singularly, attached directly to their parent thylakoid. In old or stressed chloroplasts, plastoglobuli tend to occur in linked groups or chains, still always anchored to a thylakoid.

Plastoglobuli form when a bubble appears between the layers of the lipid bilayer of the thylakoid membrane, or bud from existing plastoglubuli—though they never detach and float off into the stroma. Practically all plastoglobuli form on or near the highly curved edges of the thylakoid disks or sheets. They are also more common on stromal thylakoids than on granal ones.

Starch Granules

Starch granules are very common in chloroplasts, typically taking up 15% of the organelle's volume, though in some other plastids like amyloplasts, they can be big enough to distort the shape of the organelle. Starch granules are simply accumulations of starch in the stroma, and are not bounded by a membrane.

Starch granules appear and grow throughout the day, as the chloroplast synthesizes sugars, and are consumed at night to fuel respiration and continue sugar export into the phloem, though in mature chloroplasts, it is rare for a starch granule to be completely consumed or for a new granule to accumulate.

Starch granules vary in composition and location across different chloroplast lineages. In red algae, starch granules are found in the cytoplasm rather than in the chloroplast. In C_4 plants, mesophyll chloroplasts, which do not synthesize sugars, lack starch granules.

Rubisco

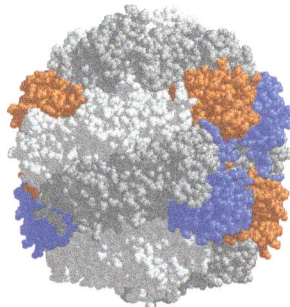

Rubisco, shown here in a space-filling model, is the main enzyme responsible for carbon fixation in chloroplasts.

The chloroplast stroma contains many proteins, though the most common and important is Rubisco, which is probably also the most abundant protein on the planet. Rubisco is the enzyme that fixes CO_2 into sugar molecules. In C_3 plants, rubisco is abundant in all chloroplasts, though in C_4 plants, it is confined to the bundle sheath chloroplasts, where the Calvin cycle is carried out in C_4 plants.

Pyrenoids

The chloroplasts of some hornworts and algae contain structures called pyrenoids. They are not found in higher plants. Pyrenoids are roughly spherical and highly refractive bodies which are a site of starch accumulation in plants that contain them. They consist of a matrix opaque to electrons, surrounded by two hemispherical starch plates. The starch is accumulated as the pyrenoids mature. In algae with carbon concentrating mechanisms, the enzyme rubisco is found in the pyrenoids. Starch can also accumulate around the pyrenoids when CO_2 is scarce. Pyrenoids can divide to form new pyrenoids, or be produced "de novo".

Thylakoid System

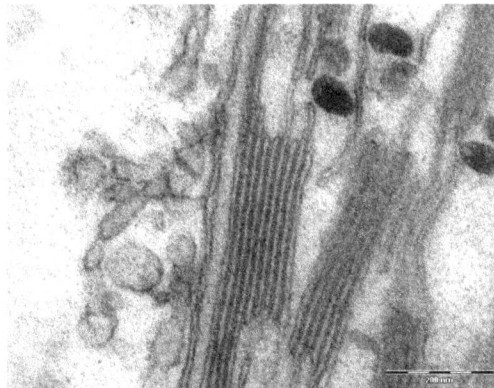

Transmission electron microscope image of some thylakoids arranged in grana stacks and lamellæ. Plastoglobuli (dark blobs) are also present.

Suspended within the chloroplast stroma is the thylakoid system, a highly dynamic collection of membranous sacks called thylakoids where chlorophyll is found and the light reactions of photosynthesis happen. In most vascular plant chloroplasts, the thylakoids are arranged in stacks called grana, though in certain C_4 plant chloroplasts and some algal chloroplasts, the thylakoids are free floating.

Granal Structure

Using a light microscope, it is just barely possible to see tiny green granules—which were named grana. With electron microscopy, it became possible to see the thylakoid system in more detail, revealing it to consist of stacks of flat thylakoids which made up the grana, and long interconnecting stromal thylakoids which linked different grana. In the transmission electron microscope, thylakoid membranes appear as alternating light-and-dark bands, 8.5 nanometers thick.

For a long time, the three-dimensional structure of the thylakoid system has been unknown or disputed. One model has the granum as a stack of thylakoids linked by helical stromal thylakoids;

the other has the granum as a single folded thylakoid connected in a "hub and spoke" way to other grana by stromal thylakoids. While the thylakoid system is still commonly depicted according to the folded thylakoid model, it was determined in 2011 that the stacked and helical thylakoids model is correct.

Granum structure The prevailing model for granal structure is a stack of granal thylakoids linked by helical stromal thylakoids that wrap around the grana stacks and form large sheets that connect different grana.

In the helical thylakoid model, grana consist of a stack of flattened circular granal thylakoids that resemble pancakes. Each granum can contain anywhere from two to a hundred thylakoids, though grana with 10–20 thylakoids are most common. Wrapped around the grana are helicoid stromal thylakoids, also known as frets or lamellar thylakoids. The helices ascend at an angle of 20–25°, connecting to each granal thylakoid at a bridge-like slit junction. The helicoids may extend as large sheets that link multiple grana, or narrow to tube-like bridges between grana. While different parts of the thylakoid system contain different membrane proteins, the thylakoid membranes are continuous and the thylakoid space they enclose form a single continuous labyrinth.

Thylakoids

Thylakoids (sometimes spelled *thylakoïds*), are small interconnected sacks which contain the membranes that the light reactions of photosynthesis take place on. The word *thylakoid* comes from the Greek word *thylakos* which means "sack".

Embedded in the thylakoid membranes are important protein complexes which carry out the light reactions of photosynthesis. Photosystem II and photosystem I contain light-harvesting complexes with chlorophyll and carotenoids that absorb light energy and use it to energize electrons. Molecules in the thylakoid membrane use the energized electrons to pump hydrogen ions into the thylakoid space, decreasing the pH and turning it acidic. ATP synthase is a large protein complex that harnesses the concentration gradient of the hydrogen ions in the thylakoid space to generate ATP energy as the hydrogen ions flow back out into the stroma—much like a dam turbine.

There are two types of thylakoids—granal thylakoids, which are arranged in grana, and stromal thylakoids, which are in contact with the stroma. Granal thylakoids are pancake-shaped circular disks about 300–600 nanometers in diameter. Stromal thylakoids are helicoid sheets that spiral around grana. The flat tops and bottoms of granal thylakoids contain only the relatively flat photosystem II protein complex. This allows them to stack tightly, forming grana with many layers of tightly appressed membrane, called granal membrane, increasing stability and surface area for light capture.

In contrast, photosystem I and ATP synthase are large protein complexes which jut out into the stroma. They can't fit in the appressed granal membranes, and so are found in the stromal thylakoid membrane—the edges of the granal thylakoid disks and the stromal thylakoids. These large protein complexes may act as spacers between the sheets of stromal thylakoids.

The number of thylakoids and the total thylakoid area of a chloroplast is influenced by light exposure. Shaded chloroplasts contain larger and more grana with more thylakoid membrane area than chloroplasts exposed to bright light, which have smaller and fewer grana and less thylakoid area. Thylakoid extent can change within minutes of light exposure or removal.

Pigments and Chloroplast Colors

Inside the photosystems embedded in chloroplast thylakoid membranes are various photosynthetic pigments, which absorb and transfer light energy. The types of pigments found are different in various groups of chloroplasts, and are responsible for a wide variety of chloroplast colorations.

Chlorophylls

Chlorophyll *a* is found in all chloroplasts, as well as their cyanobacterial ancestors. Chlorophyll *a* is a blue-green pigment partially responsible for giving most cyanobacteria and chloroplasts their color. Other forms of chlorophyll exist, such as the accessory pigments chlorophyll *b*, chlorophyll *c*, chlorophyll *d*, and chlorophyll *f*.

Chlorophyll *b* is an olive green pigment found only in the chloroplasts of plants, green algae, any secondary chloroplasts obtained through the secondary endosymbiosis of a green alga, and a few cyanobacteria. It is the chlorophylls *a* and *b* together that make most plant and green algal chloroplasts green.

Chlorophyll *c* is mainly found in secondary endosymbiotic chloroplasts that originated from a red alga, although it is not found in chloroplasts of red algae themselves. Chlorophyll *c* is also found in some green algae and cyanobacteria.

Chlorophylls *d* and *f* are pigments found only in some cyanobacteria.

Carotenoids

Delesseria sanguinea, a red alga, has chloroplasts that contain red pigments like phycoerytherin that mask their blue-green chlorophyll *a*.

In addition to chlorophylls, another group of yellow–orange pigments called carotenoids are also found in the photosystems. There are about thirty photosynthetic carotenoids. They help transfer and dissipate excess energy, and their bright colors sometimes override the chlorophyll green, like during the fall, when the leaves of some land plants change color. β-carotene is a bright red-orange carotenoid found in nearly all chloroplasts, like chlorophyll *a*. Xanthophylls, especially the orange-red zeaxanthin, are also common. Many other forms of carotenoids exist that are only found in certain groups of chloroplasts.

Phycobilins

Phycobilins are a third group of pigments found in cyanobacteria, and glaucophyte, red algal, and cryptophyte chloroplasts. Phycobilins come in all colors, though phycoerytherin is one of the pigments that makes many red algae red. Phycobilins often organize into relatively large protein complexes about 40 nanometers across called phycobilisomes. Like photosystem I and ATP synthase, phycobilisomes jut into the stroma, preventing thylakoid stacking in red algal chloroplasts. Cryptophyte chloroplasts and some cyanobacteria don't have their phycobilin pigments organized into phycobilisomes, and keep them in their thylakoid space instead.

Specialized Chloroplasts in C4 Plants

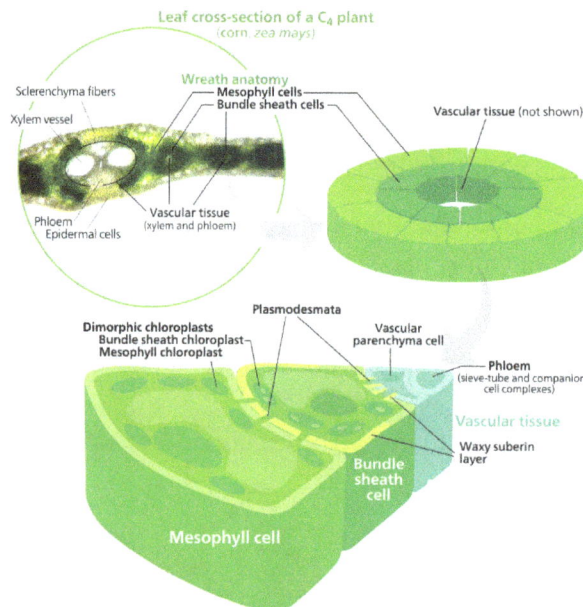

Many C_4 plants have their mesophyll cells and bundle sheath cells arranged radially around their leaf veins. The two types of cells contain different types of chloroplasts specialized for a particular part of photosynthesis.

To fix carbon dioxide into sugar molecules in the process of photosynthesis, chloroplasts use an enzyme called rubisco. Rubisco has a problem—it has trouble distinguishing between carbon dioxide and oxygen, so at high oxygen concentrations, rubisco starts accidentally adding oxygen to sugar precursors. This has the end result of ATP energy being wasted and CO_2 being released, all with no sugar being produced. This is a big problem, since O_2 is produced by the initial light reactions of photosynthesis, causing issues down the line in the Calvin cycle which uses rubisco.

C_4 plants evolved a way to solve this—by spatially separating the light reactions and the Calvin cycle. The light reactions, which store light energy in ATP and NADPH, are done in the mesophyll cells of a C_4 leaf. The Calvin cycle, which uses the stored energy to make sugar using rubisco, is done in the bundle sheath cells, a layer of cells surrounding a vein in a leaf.

As a result, chloroplasts in C_4 mesophyll cells and bundle sheath cells are specialized for each stage of photosynthesis. In mesophyll cells, chloroplasts are specialized for the light reactions, so they lack rubisco, and have normal grana and thylakoids, which they use to make ATP and NADPH, as well as oxygen. They store CO_2 in a four-carbon compound, which is why the process is called C_4 *photosynthesis*. The four-carbon compound is then transported to the bundle sheath chloroplasts, where it drops off CO_2 and returns to the mesophyll. Bundle sheath chloroplasts do not carry out the light reactions, preventing oxygen from building up in them and disrupting rubisco activity. Because of this, they lack thylakoids organized into grana stacks—though bundle sheath chloroplasts still have free-floating thylakoids in the stroma where they still carry out cyclic electron flow, a light-driven method of synthesizing ATP to power the Calvin cycle without generating oxygen. They lack photosystem II, and only have photosystem I—the only protein complex needed for cyclic electron flow. Because the job of bundle sheath chloroplasts is to carry out the Calvin cycle and make sugar, they often contain large starch grains.

Both types of chloroplast contain large amounts of chloroplast peripheral reticulum, which they use to get more surface area to transport stuff in and out of them. Mesophyll chloroplasts have a little more peripheral reticulum than bundle sheath chloroplasts.

Location

Distribution in a Plant

Not all cells in a multicellular plant contain chloroplasts. All green parts of a plant contain chloroplasts—the chloroplasts, or more specifically, the chlorophyll in them are what make the photosynthetic parts of a plant green. The plant cells which contain chloroplasts are usually parenchyma cells, though chloroplasts can also be found in collenchyma tissue. A plant cell which contains chloroplasts is known as a chlorenchyma cell. A typical chlorenchyma cell of a land plant contains about 10 to 100 chloroplasts.

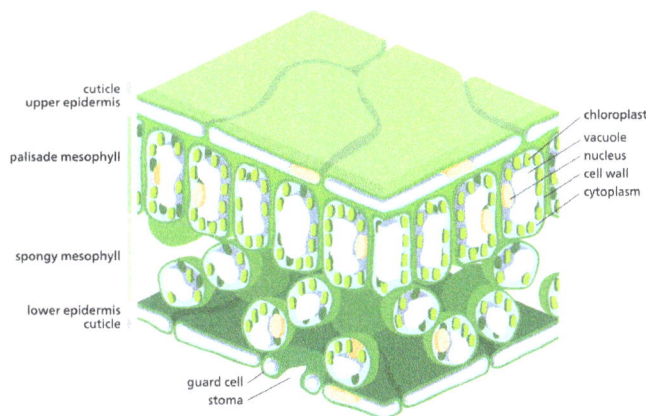

A cross section of a leaf, showing chloroplasts in its mesophyll cells. Stomal guard cells also have chloroplasts, though much fewer than mesophyll cells.

In some plants such as cacti, chloroplasts are found in the stems, though in most plants, chloroplasts are concentrated in the leaves. One square millimeter of leaf tissue can contain half a million chloroplasts. Within a leaf, chloroplasts are mainly found in the mesophyll layers of a leaf, and the guard cells of stomata. Palisade mesophyll cells can contain 30–70 chloroplasts per cell, while stomatal guard cells contain only around 8–15 per cell, as well as much less chlorophyll. Chloroplasts can also be found in the bundle sheath cells of a leaf, especially in C_4 plants, which carry out the Calvin cycle in their bundle sheath cells. They are often absent from the epidermis of a leaf.

Cellular Location

Chloroplast Movement

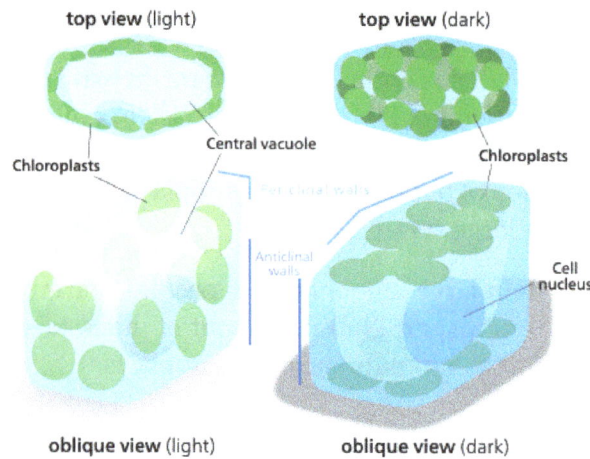

top view (light)   top view (dark)

Chloroplasts   Central vacuole   Chloroplasts

                          Periclinal walls

                          Anticlinal   Cell
                          walls   nucleus

oblique view (light)   oblique view (dark)

When chloroplasts are exposed to direct sunlight, they stack along the anticlinal cell walls
to minimize exposure. In the dark they spread out in sheets along the
periclinal walls to maximize light absorption.

The chloroplasts of plant and algal cells can orient themselves to best suit the available light. In low-light conditions, they will spread out in a sheet—maximizing the surface area to absorb light. Under intense light, they will seek shelter by aligning in vertical columns along the plant cell's cell wall or turning sideways so that light strikes them edge-on. This reduces exposure and protects them from photooxidative damage. This ability to distribute chloroplasts so that they can take shelter behind each other or spread out may be the reason why land plants evolved to have many small chloroplasts instead of a few big ones. Chloroplast movement is considered one of the most closely regulated stimulus-response systems that can be found in plants. Mitochondria have also been observed to follow chloroplasts as they move.

In higher plants, chloroplast movement is run by phototropins, blue light photoreceptors also responsible for plant phototropism. In some algae, mosses, ferns, and flowering plants, chloroplast movement is influenced by red light in addition to blue light, though very long red wavelengths inhibit movement rather than speeding it up. Blue light generally causes chloroplasts to seek shelter, while red light draws them out to maximize light absorption.

Studies of *Vallisneria gigantea*, an aquatic flowering plant, have shown that chloroplasts can get moving within five minutes of light exposure, though they don't initially show any net directionality. They may move along microfilament tracks, and the fact that the microfilament mesh changes shape to form a honeycomb structure surrounding the chloroplasts after they have moved suggests that microfilaments may help to anchor chloroplasts in place.

Function and Chemistry

Guard Cell Chloroplasts

Unlike most epidermal cells, the guard cells of plant stomata contain relatively well-developed chloroplasts. However, exactly what they do is controversial.

Plant Innate Immunity

Plants lack specialized immune cells—all plant cells participate in the plant immune response. Chloroplasts, along with the nucleus, cell membrane, and endoplasmic reticulum, are key players in pathogen defense. Due to its role in a plant cell's immune response, pathogens frequently target the chloroplast.

Plants have two main immune responses—the hypersensitive response, in which infected cells seal themselves off and undergo programmed cell death, and systemic acquired resistance, where infected cells release signals warning the rest of the plant of a pathogen's presence. Chloroplasts stimulate both responses by purposely damaging their photosynthetic system, producing reactive oxygen species. High levels of reactive oxygen species will cause the hypersensitive response. The reactive oxygen species also directly kill any pathogens within the cell. Lower levels of reactive oxygen species initiate systemic acquired resistance, triggering defense-molecule production in the rest of the plant.

In some plants, chloroplasts are known to move closer to the infection site and the nucleus during an infection.

Chloroplasts can serve as cellular sensors. After detecting stress in a cell, which might be due to a pathogen, chloroplasts begin producing molecules like salicylic acid, jasmonic acid, nitric oxide and reactive oxygen species which can serve as defense-signals. As cellular signals, reactive oxygen species are unstable molecules, so they probably don't leave the chloroplast, but instead pass on their signal to an unknown second messenger molecule. All these molecules initiate retrograde signaling—signals from the chloroplast that regulate gene expression in the nucleus.

In addition to defense signaling, chloroplasts, with the help of the peroxisomes, help synthesize an important defense molecule, jasmonate. Chloroplasts synthesize all the fatty acids in a plant cell—linoleic acid, a fatty acid, is a precursor to jasmonate.

Photosynthesis

One of the main functions of the chloroplast is its role in photosynthesis, the process by which light is transformed into chemical energy, to subsequently produce food in the form of sugars. Water (H_2O) and carbon dioxide (CO_2) are used in photosynthesis, and sugar and oxygen (O_2) is made, using light energy. Photosynthesis is divided into two stages—the light reactions, where water is split to produce oxygen, and the dark reactions, or Calvin cycle, which builds sugar molecules from carbon dioxide. The two phases are linked by the energy carriers adenosine triphosphate (ATP) and nicotinamide adenine dinucleotide phosphate ($NADP^+$).

Light Reactions

The light reactions of photosynthesis take place across the thylakoid membranes.

Energy Carriers

ATP is the phosphorylated version of adenosine diphosphate (ADP), which stores energy in a cell and powers most cellular activities. ATP is the energized form, while ADP is the (partially) depleted form. $NADP^+$ is an electron carrier which ferries high energy electrons. In the light reactions, it gets reduced, meaning it picks up electrons, becoming NADPH.

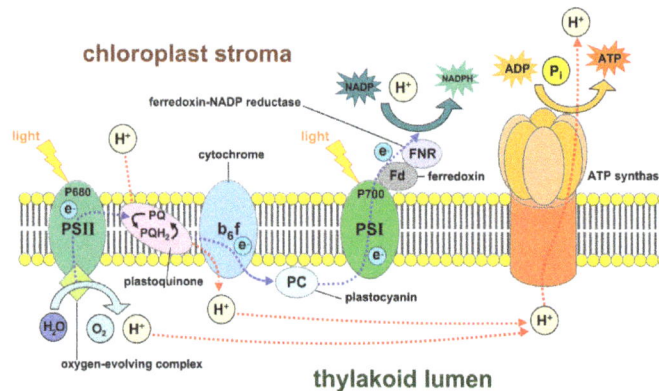

The light reactions take place on the thylakoid membranes. They take light energy and store it in NADPH, a form of $NADP^+$, and ATP to fuel the dark reactions.

Photophosphorylation

Like mitochondria, chloroplasts use the potential energy stored in an H^+, or hydrogen ion gradient to generate ATP energy. The two photosystems capture light energy to energize electrons taken from water, and release them down an electron transport chain. The molecules between the photosystems harness the electrons' energy to pump hydrogen ions into the thylakoid space, creating a concentration gradient, with more hydrogen ions (up to a thousand times as many) inside the thylakoid system than in the stroma. The hydrogen ions in the thylakoid space then diffuse back down their concentration gradient, flowing back out into the stroma through ATP synthase. ATP synthase uses the energy from the flowing hydrogen ions to phosphorylate adenosine diphosphate into adenosine triphosphate, or ATP. Because chloroplast ATP synthase projects out into the stroma, the ATP is synthesized there, in position to be used in the dark reactions.

$NADP^+$ Reduction

Electrons are often removed from the electron transport chains to charge $NADP^+$ with electrons, reducing it to NADPH. Like ATP synthase, ferredoxin-$NADP^+$ reductase, the enzyme that reduces $NADP^+$, releases the NADPH it makes into the stroma, right where it is needed for the dark reactions.

Because $NADP^+$ reduction removes electrons from the electron transport chains, they must be replaced—the job of photosystem II, which splits water molecules (H_2O) to obtain the electrons from its hydrogen atoms.

Cyclic Photophosphorylation

While photosystem II photolyzes water to obtain and energize new electrons, photosystem I simply reenergizes depleted electrons at the end of an electron transport chain. Normally, the reenergized electrons are taken by $NADP^+$, though sometimes they can flow back down more H^+-pumping electron transport chains to transport more hydrogen ions into the thylakoid space to generate more ATP. This is termed cyclic photophosphorylation because the electrons are recycled. Cyclic photophosphorylation is common in C_4 plants, which need more ATP than NADPH.

Dark Reactions

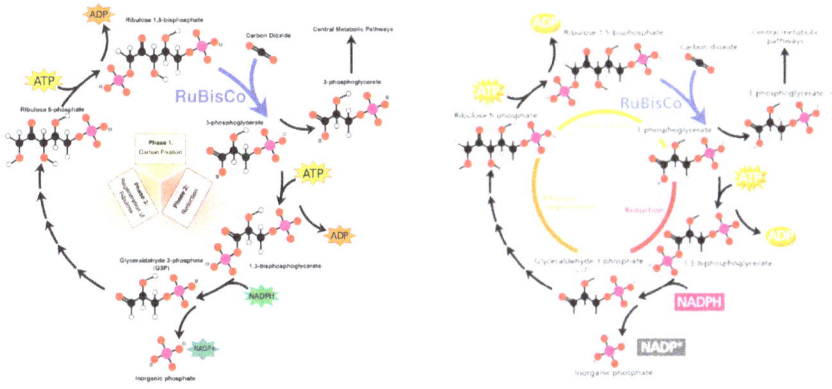

The Calvin cycle incorporates carbon dioxide into sugar molecules.

The Calvin cycle, also known as the dark reactions, is a series of biochemical reactions that fixes CO_2 into G3P sugar molecules and uses the energy and electrons from the ATP and NADPH made in the light reactions. The Calvin cycle takes place in the stroma of the chloroplast.

While named *"the dark reactions"*, in most plants, they take place in the light, since the dark reactions are dependent on the products of the light reactions.

Carbon Fixation and G3P Synthesis

The Calvin cycle starts by using the enzyme Rubisco to fix CO_2 into five-carbon Ribulose bisphosphate (RuBP) molecules. The result is unstable six-carbon molecules that immediately break down into three-carbon molecules called 3-phosphoglyceric acid, or 3-PGA. The ATP and NADPH made in the light reactions is used to convert the 3-PGA into glyceraldehyde-3-phosphate, or G3P sugar molecules. Most of the G3P molecules are recycled back into RuBP using energy from more ATP, but one out of every six produced leaves the cycle—the end product of the dark reactions.

Sugars and Starches

Glyceraldehyde-3-phosphate can double up to form larger sugar molecules like glucose and fructose. These molecules are processed, and from them, the still larger sucrose, a disaccharide commonly known as table sugar, is made, though this process takes place outside of the chloroplast, in the cytoplasm.

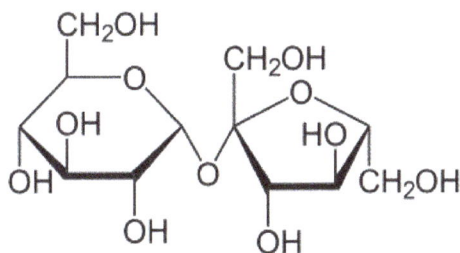

Sucrose is made up of a glucose monomer (left), and a fructose monomer (right).

Alternatively, glucose monomers in the chloroplast can be linked together to make starch, which accumulates into the starch grains found in the chloroplast. Under conditions such as high atmospheric CO_2 concentrations, these starch grains may grow very large, distorting the grana and thylakoids. The starch granules displace the thylakoids, but leave them intact. Waterlogged roots can also cause starch buildup in the chloroplasts, possibly due to less sucrose being exported out of the chloroplast (or more accurately, the plant cell). This depletes a plant's free phosphate supply, which indirectly stimulates chloroplast starch synthesis. While linked to low photosynthesis rates, the starch grains themselves may not necessarily interfere significantly with the efficiency of photosynthesis, and might simply be a side effect of another photosynthesis-depressing factor.

Photorespiration

Photorespiration can occur when the oxygen concentration is too high. Rubisco cannot distinguish between oxygen and carbon dioxide very well, so it can accidentally add O_2 instead of CO_2 to RuBP. This process reduces the efficiency of photosynthesis—it consumes ATP and oxygen, releases CO_2, and produces no sugar. It can waste up to half the carbon fixed by the Calvin cycle. Several mechanisms have evolved in different lineages that raise the carbon dioxide concentration relative to oxygen within the chloroplast, increasing the efficiency of photosynthesis. These mechanisms are called carbon dioxide concentrating mechanisms, or CCMs. These include Crassulacean acid metabolism, C_4 carbon fixation, and pyrenoids. Chloroplasts in C_4 plants are notable as they exhibit a distinct chloroplast dimorphism.

pH

Because of the H^+ gradient across the thylakoid membrane, the interior of the thylakoid is acidic, with a pH around 4, while the stroma is slightly basic, with a pH of around 8. The optimal stroma pH for the Calvin cycle is 8.1, with the reaction nearly stopping when the pH falls below 7.3.

CO_2 in water can form carbonic acid, which can disturb the pH of isolated chloroplasts, interfering with photosynthesis, even though CO_2 is used in photosynthesis. However, chloroplasts in living plant cells are not affected by this as much.

Chloroplasts can pump K^+ and H^+ ions in and out of themselves using a poorly understood light-driven transport system.

In the presence of light, the pH of the thylakoid lumen can drop up to 1.5 pH units, while the pH of the stroma can rise by nearly one pH unit.

Amino Acid Synthesis

Chloroplasts alone make almost all of a plant cell's amino acids in their stroma except the sulfur-containing ones like cysteine and methionine. Cysteine is made in the chloroplast (the proplastid too) but it is also synthesized in the cytosol and mitochondria, probably because it has trouble crossing membranes to get to where it is needed. The chloroplast is known to make the precursors to methionine but it is unclear whether the organelle carries out the last leg of the pathway or if it happens in the cytosol.

Other Nitrogen Compounds

Chloroplasts make all of a cell's purines and pyrimidines—the nitrogenous bases found in DNA and RNA. They also convert nitrite (NO_2^-) into ammonia (NH_3) which supplies the plant with nitrogen to make its amino acids and nucleotides.

Other Chemical Products

Chloroplasts are the site of complex lipid metabolism.

Differentiation, Replication, and Inheritance

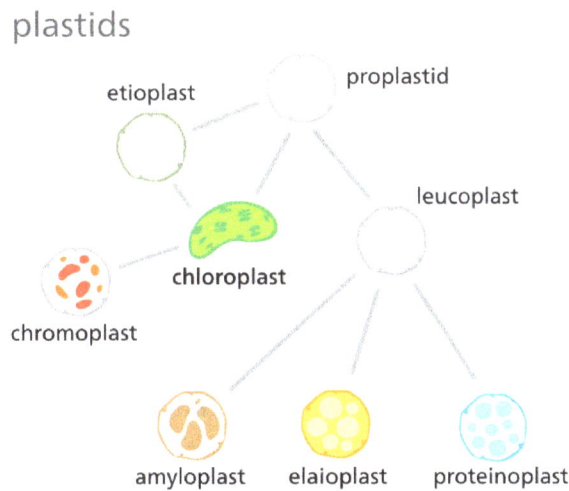

Plastid types *(Interactive diagram)* Plants contain many different kinds of plastids in their cells.

Chloroplasts are a special type of a plant cell organelle called a plastid, though the two terms are sometimes used interchangeably. There are many other types of plastids, which carry out various functions. All chloroplasts in a plant are descended from undifferentiated proplastids found in the zygote, or fertilized egg. Proplastids are commonly found in an adult plant's apical meristems. Chloroplasts do not normally develop from proplastids in root tip meristems—instead, the formation of starch-storing amyloplasts is more common.

In shoots, proplastids from shoot apical meristems can gradually develop into chloroplasts in photosynthetic leaf tissues as the leaf matures, if exposed to the required light. This process involves invaginations of the inner plastid membrane, forming sheets of membrane that project into the internal stroma. These membrane sheets then fold to form thylakoids and grana.

If angiosperm shoots are not exposed to the required light for chloroplast formation, proplastids may develop into an etioplast stage before becoming chloroplasts. An etioplast is a plastid that lacks chlorophyll, and has inner membrane invaginations that form a lattice of tubes in their stroma, called a prolamellar body. While etioplasts lack chlorophyll, they have a yellow chlorophyll precursor stocked. Within a few minutes of light exposure, the prolamellar body begins to reorganize into stacks of thylakoids, and chlorophyll starts to be produced. This process, where the etioplast becomes a chloroplast, takes several hours. Gymnosperms do not require light to form chloroplasts.

Light, however, does not guarantee that a proplastid will develop into a chloroplast. Whether a proplastid develops into a chloroplast some other kind of plastid is mostly controlled by the nucleus and is largely influenced by the kind of cell it resides in.

possible plastid interconversions

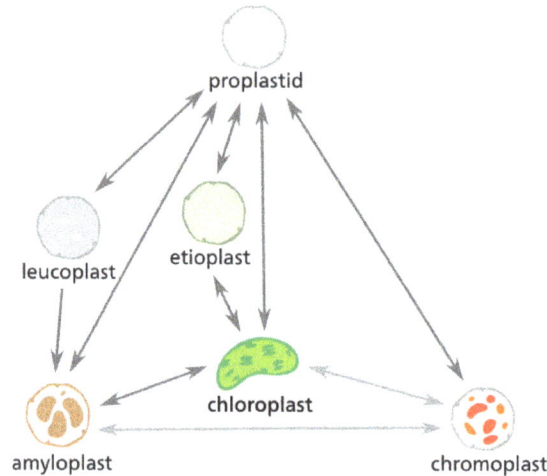

Many plastid interconversions are possible.

Plastid Interconversion

Plastid differentiation is not permanent, in fact many interconversions are possible. Chloroplasts may be converted to chromoplasts, which are pigment-filled plastids responsible for the bright colors seen in flowers and ripe fruit. Starch storing amyloplasts can also be converted to chromoplasts, and it is possible for proplastids to develop straight into chromoplasts. Chromoplasts and amyloplasts can also become chloroplasts, like what happens when a carrot or a potato is illuminated. If a plant is injured, or something else causes a plant cell to revert to a meristematic state, chloroplasts and other plastids can turn back into proplastids. Chloroplast, amyloplast, chromoplast, proplast, etc., are not absolute states—intermediate forms are common.

Chloroplast Division

Most chloroplasts in a photosynthetic cell do not develop directly from proplastids or etioplasts. In fact, a typical shoot meristematic plant cell contains only 7–20 proplastids. These proplastids differentiate into chloroplasts, which divide to create the 30–70 chloroplasts found in a mature photosynthetic plant cell. If the cell divides, chloroplast division provides the additional chloroplasts to partition between the two daughter cells.

In single-celled algae, chloroplast division is the only way new chloroplasts are formed. There is no proplastid differentiation—when an algal cell divides, its chloroplast divides along with it, and each daughter cell receives a mature chloroplast.

Almost all chloroplasts in a cell divide, rather than a small group of rapidly dividing chloroplasts. Chloroplasts have no definite S-phase—their DNA replication is not synchronized or limited to that of their host cells. Much of what we know about chloroplast division comes from studying organisms like *Arabidopsis* and the red alga *Cyanidioschyzon merolæ*.

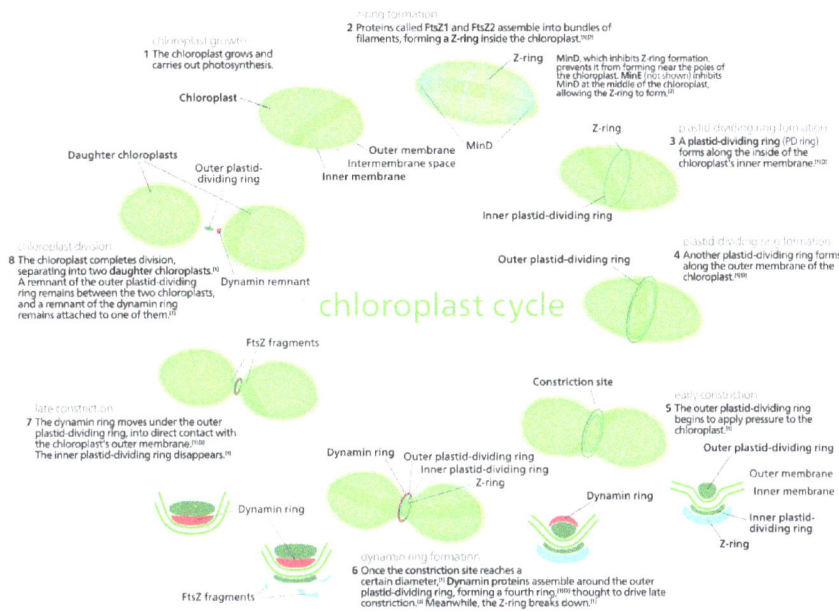

Most chloroplasts in plant cells, and all chloroplasts in algae arise from chloroplast division.

The division process starts when the proteins FtsZ1 and FtsZ2 assemble into filaments, and with the help of a protein ARC6, form a structure called a Z-ring within the chloroplast's stroma. The Min system manages the placement of the Z-ring, ensuring that the chloroplast is cleaved more or less evenly. The protein MinD prevents FtsZ from linking up and forming filaments. Another protein ARC3 may also be involved, but it is not very well understood. These proteins are active at the poles of the chloroplast, preventing Z-ring formation there, but near the center of the chloroplast, MinE inhibits them, allowing the Z-ring to form.

Next, the two plastid-dividing rings, or PD rings form. The inner plastid-dividing ring is located in the inner side of the chloroplast's inner membrane, and is formed first. The outer plastid-dividing ring is found wrapped around the outer chloroplast membrane. It consists of filaments about 5 nanometers across, arranged in rows 6.4 nanometers apart, and shrinks to squeeze the chloroplast. This is when chloroplast constriction begins. In a few species like *Cyanidioschyzon merolæ*, chloroplasts have a third plastid-dividing ring located in the chloroplast's intermembrane space.

Late into the constriction phase, dynamin proteins assemble around the outer plastid-dividing ring, helping provide force to squeeze the chloroplast. Meanwhile, the Z-ring and the inner plastid-dividing ring break down. During this stage, the many chloroplast DNA plasmids floating around in the stroma are partitioned and distributed to the two forming daughter chloroplasts.

Later, the dynamins migrate under the outer plastid dividing ring, into direct contact with the chloroplast's outer membrane, to cleave the chloroplast in two daughter chloroplasts.

A remnant of the outer plastid dividing ring remains floating between the two daughter chloroplasts, and a remnant of the dynamin ring remains attached to one of the daughter chloroplasts.

Of the five or six rings involved in chloroplast division, only the outer plastid-dividing ring is present for the entire constriction and division phase—while the Z-ring forms first, constriction does not begin until the outer plastid-dividing ring forms.

Chloroplast division In this light micrograph of some moss chloroplasts, many dumbbell-shaped chloroplasts can be seen dividing. Grana are also just barely visible as small granules.

Regulation

In species of algae that contain a single chloroplast, regulation of chloroplast division is extremely important to ensure that each daughter cell receives a chloroplast—chloroplasts can't be made from scratch. In organisms like plants, whose cells contain multiple chloroplasts, coordination is looser and less important. It is likely that chloroplast and cell division are somewhat synchronized, though the mechanisms for it are mostly unknown.

Light has been shown to be a requirement for chloroplast division. Chloroplasts can grow and progress through some of the constriction stages under poor quality green light, but are slow to complete division—they require exposure to bright white light to complete division. Spinach leaves grown under green light have been observed to contain many large dumbbell-shaped chloroplasts. Exposure to white light can stimulate these chloroplasts to divide and reduce the population of dumbbell-shaped chloroplasts.

Chloroplast Inheritance

Like mitochondria, chloroplasts are usually inherited from a single parent. Biparental chloroplast inheritance—where plastid genes are inherited from both parent plants—occurs in very low levels in some flowering plants.

Many mechanisms prevent biparental chloroplast DNA inheritance, including selective destruction of chloroplasts or their genes within the gamete or zygote, and chloroplasts from one parent being excluded from the embryo. Parental chloroplasts can be sorted so that only one type is present in each offspring.

Gymnosperms, such as pine trees, mostly pass on chloroplasts paternally, while flowering plants often inherit chloroplasts maternally. Flowering plants were once thought to only inherit chloroplasts maternally. However, there are now many documented cases of angiosperms inheriting chloroplasts paternally.

Angiosperms, which pass on chloroplasts maternally, have many ways to prevent paternal inheritance. Most of them produce sperm cells that do not contain any plastids. There are many other documented mechanisms that prevent paternal inheritance in these flowering plants, such as different rates of chloroplast replication within the embryo.

Among angiosperms, paternal chloroplast inheritance is observed more often in hybrids than in offspring from parents of the same species. This suggests that incompatible hybrid genes might interfere with the mechanisms that prevent paternal inheritance.

Transplastomic Plants

Recently, chloroplasts have caught attention by developers of genetically modified crops. Since, in most flowering plants, chloroplasts are not inherited from the male parent, transgenes in these plastids cannot be disseminated by pollen. This makes plastid transformation a valuable tool for the creation and cultivation of genetically modified plants that are biologically contained, thus posing significantly lower environmental risks. This biological containment strategy is therefore suitable for establishing the coexistence of conventional and organic agriculture. While the reliability of this mechanism has not yet been studied for all relevant crop species, recent results in tobacco plants are promising, showing a failed containment rate of transplastomic plants at 3 in 1,000,000.

Chromoplast

Chromoplasts are plastids, heterogeneous organelles responsible for pigment synthesis and storage in specific photosynthetic eukaryotes. It is thought that like all other plastids including chloroplasts and leucoplasts they are descended from symbiotic prokaryotes.

The coloration of the petals and sepals on the Bee orchid is
controlled by a specialized organelle in plant cells called a chromoplast.

Function

Chromoplasts are found in fruits, flowers, roots, and stressed and aging leaves, and are responsible for their distinctive colors. This is always associated with a massive increase in the accumulation of carotenoid pigments. The conversion of chloroplasts to chromoplasts in ripening is a classic example.

They are generally found in mature tissues and are derived from preexisting mature plastids. Fruits and flowers are the most common structures for the biosynthesis of carotenoids, although other reactions occur there as well including the synthesis of sugars, starches, lipids, aromatic compounds, vitamins and hormones. The DNA in chloroplasts and chromoplasts is identical. One subtle difference in DNA was found after a liquid chromatography analysis of tomato chromoplasts was conducted, revealing increased cytosine methylation.

Chromoplasts synthesize and store pigments such as orange carotene, yellow xanthophylls, and various other red pigments. As such, their color varies depending on what pigment they contain. The main evolutionary purpose of chromoplasts is probably to attract pollinators or eaters of colored fruits, which help disperse seeds. However, they are also found in roots such as carrots and sweet potatoes. They allow the accumulation of large quantities of water-insoluble compounds in otherwise watery parts of plants.

When leaves change color in the autumn, it is due to the loss of green chlorophyll, which unmasks preexisting carotenoids. In this case, relatively little new carotenoid is produced—the change in plastid pigments associated with leaf senescence is somewhat different from the active conversion to chromoplasts observed in fruit and flowers.

There are some species of flowering plants that contain little to no carotenoids. In such cases there are plastids present within the petals that closely resemble chromoplasts and are sometimes visually indistinguishable. Anthocyanins and flavonoids located in the cell vacuoles are responsible for other colors of pigment.

The term "chromoplast" is occasionally used to include *any* plastid that has pigment, mostly to emphasize the difference between them and the various types of leucoplasts, plastids that have no pigments. In this sense, chloroplasts are a specific type of chromoplast. Still, "chromoplast" is more often used to denote plastids with pigments other than chlorophyll.

Structure and Classification

Using a light microscope chromoplasts can be differentiated and are classified into four main types. The first type is composed of proteic stroma with granules. The second is composed of protein crystals and amorphous pigment granules. The third type is composed of protein and pigment crystals. The fourth type is a chromoplast which only contains crystals. An electron microscope reveals even more, allowing for the identification of substructures such as globules, crystals, membranes, fibrils and tubules. The substructures found in chromoplasts are not found in the mature plastid that it divided from.

The presence, frequency and identification of substructures using an electron microscope has led to further classification, dividing chromoplasts into five main categories: Globular chromoplasts, crystalline chromoplasts, fibrillar chromoplasts, tubular chromoplasts and membranous chromoplasts. It has also been found that different types of chromoplasts can coexist in the same organ. Some examples of plants in the various categories include mangos, which have globular chromoplasts, and carrots which have crystalline chromoplasts.

Although some chromoplasts are easily categorized, others have characteristics from multiple categories that make them hard to place. Tomatoes accumulate carotenoids, mainly lycopene crystalloids in membrane-shaped structures, which could place them in either the crystalline or membranous category.

Evolution

Plastids are descendants of cyanobacteria, photosynthetic prokaryotes, which integrated themselves into the eukaryotic ancestor of algæ and plants, forming an endosymbiotic relationship. The ancestors of plastids diversified into a variety of plastid types, including chromoplasts. Plastids also possess their own small genome and some have the ability to produce a percentage of their own proteins.

The main evolutionary purpose of chromoplasts is to attract animals and insects to pollinate their flowers and disperse their seeds. The bright colors often produced by chromoplasts is one of many ways to achieve this. Many plants have evolved symbiotic relationships with a single pollinator. Color can be a very important factor in determining which pollinators visit a flower, as specific colors attract specific pollinators. White flowers tend to attract beetles, bees are most often attracted to violet and blue flowers, and butterflies are often attracted to warmer colors like yellows and oranges.

Research

Chromoplasts are not widely studied and are rarely the main focus of scientific research. They often play a role in research on the tomato plant (*Solanum lycopersicum*). Lycopene is responsible for the red color of a ripe fruit in the cultivated tomato, while the yellow color of the flowers is due to xanthophylls violaxanthin and neoxanthin.

Carotenoid biosynthesis occurs in both chromoplasts and chloroplasts. In the chromoplasts of tomato flowers, carotenoid synthesis is regulated by the genes Psyl, Pds, Lcy-b, and Cyc-b. These genes, in addition to others, are responsible for the formation of carotenoids in organs and structures. For example, the Lcy-e gene is highly expressed in leaves, which results in the production of the carotenoid lutein.

White flowers are caused by a recessive allele in tomato plants. They are less desirable in cultivated crops because they have a lower pollination rate. In one study, it was found that chromoplasts are still present in white flowers. The lack of yellow pigment in their petals and anthers is due to a mutation in the CrtR-b2 gene which disrupts the carotenoid biosynthesis pathway.

The entire process of chromoplast formation is not yet completely understood on the molecular level. However, electron microscopy has revealed part of the transformation from chloroplast to chromoplast. The transformation starts with remodeling of the internal membrane system with the lysis of the intergranal thylakoids and the grana. New membrane systems form in organized membrane complexes called thylakoid plexus. The new membranes are the site of the formation of carotenoid crystals. These newly synthesized membranes do not come from the thylakoids, but rather from vesicles generated from the inner membrane of the plastid. The most obvious biochemical change would be the downregulation of photosynthetic gene expression which results in the loss of chlorophyll and stops photosynthetic activity.

In oranges, the synthesis of carotenoids and the disappearance of chlorophyll causes the color of the fruit to change from green to yellow. The orange color is often added artificially—light yellow-orange is the natural color created by the actual chromoplasts.

Valencia oranges *Citris sinensis L* are a cultivated orange grown extensively in the state of Florida. In the winter, Valencia oranges reach their optimum orange-rind color while reverting to a green color in the spring and summer. While it was originally thought that chromoplasts were the final stage of plastid development, in 1966 it was proved that chromoplasts can revert to chloroplasts, which causes the oranges to turn back to green.

Compare

- Plastid

    o Chloroplast and etioplast

    o Chromoplast

- ○ Leucoplast

  - ▪ Amyloplast

  - ▪ Elaioplast

  - ▪ Proteinoplast

Gerontoplast

A gerontoplast is a plastid found in formerly green tissues that are currently senescing. A gerontoplast is a chloroplast that has re-purposed through a process of developmental senescence.

Transformation of Chloroplasts to Gerontoplasts

The term *gerontoplast* was first introduced by Sitte (1977) to define the unique features of the plastid formed during leaf senescence. The process of senescence brings about regulated dismantling of cellular organelles. Chloroplast shows the first sign of senescence induced degradation and is the last organelle to survive when other organelles are completely disorganized. The formation of gerontoplast from chloroplast during senescence involves extensive structural modifications of thylakoid membrane with the concomitant formation of a large number of plastoglobuli with lipophilic materials. The envelope of the plastid, however, remains intact.

Leucoplast

Leucoplasts (leukós "white", plastós "formed, molded") are a category of plastid and as such are organelles found in plant cells. They are non-pigmented, in contrast to other plastids such as the chloroplast.

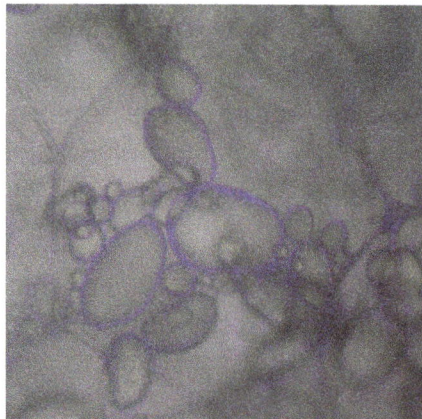

Leucoplasts, specifically, amyloplasts

Lacking photosynthetic pigments, leucoplasts are not green and are located in non-photosynthetic tissues of plants, such as roots, bulbs and seeds. They may be specialized for bulk storage of starch, lipid or protein and are then known as amyloplasts, elaioplasts, or proteinoplasts (also called aleuroplasts) respectively. However, in many cell types, leucoplasts do not have a major storage function and are present to provide a wide range of essential biosynthetic functions, including the synthesis of fatty acids such as palmitic acid, many amino acids, and tetrapyrrole compounds such as heme. In general, leucoplasts are much smaller than chloroplasts and have a variable morphology, often described as amoeboid. Extensive networks of stromules interconnecting leucoplasts have been observed in epidermal cells of roots, hypocotyls, and petals, and in callus and suspension culture cells of tobacco. In some cell types at certain stages of development, leucoplasts are clustered around the nucleus with stromules extending to the cell periphery, as observed for proplastids in the root meristem.

Etioplasts, which are pre-granal, immature chloroplasts but can also be chloroplasts that have been deprived of light, lack active pigment and can be considered leucoplasts. After several minutes exposure to light, etioplasts begin to transform into functioning chloroplasts and cease being leucoplasts. Amyloplasts are of large size and store starch. Proteinoplasts store proteins and are found in seeds (pulses). Elaioplasts store fats and oils and are found in seeds. They are also called oleosomes. [castor, groundnut] Etioplasts are plastids without pigments and store food and lamellar structures. These plastids occur in etiolated plants due to the absence of light.

Compare

- Plastid

  o Chloroplast and etioplast

  o Chromoplast

  o Tannosome

  o Leucoplast

    - Amyloplast

    - Elaioplast

    - Proteinoplast

References

- Zhang, Q.; Sodmergen (2010). "Why does biparental plastid inheritance revive in angiosperms?". Journal of Plant Research. 123 (2): 201–206. PMID 20052516. doi:10.1007/s10265-009-0291-z

- Jones, Daniel (2003) [1917], Peter Roach, James Hartmann and Jane Setter, eds., English Pronouncing Dictionary, Cambridge: Cambridge University Press, ISBN 3-12-539683-2

- Milo, Ron; Philips, Rob. "Cell Biology by the Numbers: How large are chloroplasts?". book.bionumbers.org. Retrieved 7 February 2017

- Archibald, John M. (2009). "The Puzzle of Plastid Evolution". Current Biology. 19 (2): R81–8. PMID 19174147. doi:10.1016/j.cub.2008.11.067

- Keeling, Patrick J. (2004). "Diversity and evolutionary history of plastids and their hosts". American Journal of Botany. 91 (10): 1481–93. PMID 21652304. doi:10.3732/ajb.91.10.1481

- Wise, edited by Robert R.; Hoober, J. Kenneth (2006). The structure and function of plastids. Dordrecht: Springer. pp. 3–21. ISBN 978-1-4020-4061-0

- Schnepf, Eberhard; Elbrächter, Malte (1999). "Dinophyte chloroplasts and phylogeny – A review". Grana. 38 (2–3): 81–97. doi:10.1080/00173139908559217

- Nair, Sethu C.; Striepen, Boris (2011). "What Do Human Parasites Do with a Chloroplast Anyway?". PLoS Biology. 9 (8): e1001137. PMC 3166169. PMID 21912515. doi:10.1371/journal.pbio.1001137

- John, D.M.; Brook, A.J.; Whitton, B.A. (2002). The freshwater algal flora of the British Isles : an identification guide to freshwater and terrestrial algae. Cambridge: Cambridge University Press. p. 335. ISBN 978-0-521-77051-4

- Eckardt, N. A. (2003). "Controlling Organelle Positioning: A Novel Chloroplast Movement Protein". The Plant Cell Online. 15 (12): 2755–7. PMC 540263. doi:10.1105/tpc.151210

- Soll, J r.; Soll, J (1996). "Phosphorylation of the Transit Sequence of Chloroplast Precursor Proteins". Journal of Biological Chemistry. 271 (11): 6545–54. PMID 8626459. doi:10.1074/jbc.271.11.6545

- Werdan, Karl; Heldt, Hans W.; Milovancev, Mirjana (1975). "The role of pH in the regulation of carbon fixation in the chloroplast stroma. Studies on CO_2 fixation in the light and dark". Biochimica et Biophysica Acta (BBA) – Bioenergetics. 396 (2): 276–92. PMID 239746. doi:10.1016/0005-2728(75)90041-9

- Burgess,, Jeremy (1989). An introduction to plant cell development (Pbk. ed.). Cambridge: Cambridge university press. p. 46. ISBN 0-521-31611-1

- Retallack, B; Butler, RD (1970). "The development and structure of pyrenoids in Bulbochaete hiloensis". Journal of Cell Science. 6 (1): 229–41. PMID 5417694

- Berg JM; Tymoczko JL; Stryer L. (2002). Biochemistry. (5th ed.). W H Freeman. pp. Section 19.4. Retrieved 30 October 2012

- J D, Rochaix (1998). The molecular biology of chloroplasts and mitochondria in Chlamydomonas. Dordrecht [u.a.]: Kluwer Acad. Publ. pp. 550–565. ISBN 978-0-7923-5174-0

- Fuks, Bruno; Fabrice Homblé (8 July 1996). "Mechanism of Proton Permeation through Chloroplast Lipid Membranes" (PDF). Plant Physiology. 112 (2): 759–766. PMC 158000. PMID 8883387. doi:10.1104/pp.112.2.759. Retrieved 25 August 2013

- Archibald, John M. (2006). "Algal Genomics: Exploring the Imprint of Endosymbiosis". Current Biology. 16 (24): R1033–5. PMID 17174910. doi:10.1016/j.cub.2006.11.008

- Takagi, Shingo (December 2002). "Actin-based photo-orientation movement of chloroplasts in plant cells". Journal of Experimental Biology. 206: 1963–1969. doi:10.1242/jeb.00215. Retrieved 6 March 2013

- Steer, Brian E.S. Gunning, Martin W. (1996). Plant cell biology : structure and function. Boston, Mass.: Jones and Bartlett Publishers. p. 24. ISBN 0-86720-504-0

Structural Division of Plant Anatomy

Plants can be structurally divided into flowers, leaves, bark and seeds. A flower can further be divided into petals, sepal, gynoecium and stamen whereas a bark constitutes of phloem, cork cambium, cortex, vascular cambium and root. This chapter discusses the methods of plant anatomy in a critical manner providing key analysis to the subject matter.

Flower

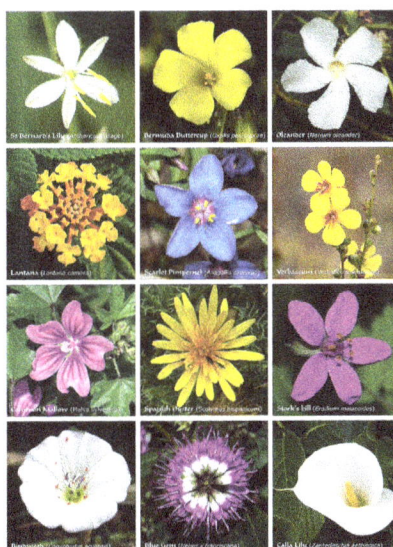

A poster with flowers or clusters of flowers produced by twelve species of flowering plants from different families

A flower, sometimes known as a bloom or blossom, is the reproductive structure found in plants that are floral (plants of the division Magnoliophyta, also called angiosperms). The biological function of a flower is to effect reproduction, usually by providing a mechanism for the union of sperm with eggs. Flowers may facilitate outcrossing (fusion of sperm and eggs from different individuals in a population) or allow selfing (fusion of sperm and egg from the same flower). Some flowers produce diaspores without fertilization (parthenocarpy). Flowers contain sporangia and are the site where gametophytes develop. Many flowers have evolved to be attractive to animals, so as to cause them to be vectors for the transfer of pollen. After fertilization, the ovary of the flower develops into fruit containing seeds.

In addition to facilitating the reproduction of flowering plants, flowers have long been admired and used by humans to bring beauty to their environment, and also as objects of romance, ritual, religion, medicine and as a source of food.

Morphology

Main parts of a mature flower (Ranunculus glaberrimus)

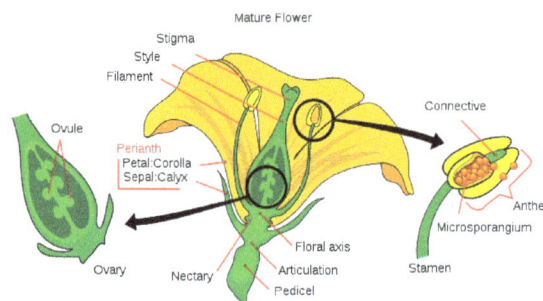

Diagram of flower parts

Floral Parts

The essential parts of a flower can be considered in two parts: the vegetative part, consisting of petals and associated structures in the perianth, and the reproductive or sexual parts. A stereotypical flower consists of four kinds of structures attached to the tip of a short stalk. Each of these kinds of parts is arranged in a whorl on the receptacle. The four main whorls (starting from the base of the flower or lowest node and working upwards) are as follows:

Perianth

Collectively the calyx and corolla form the perianth.

- *Calyx*: the outermost whorl consisting of units called *sepals*; these are typically green and enclose the rest of the flower in the bud stage, however, they can be absent or prominent and petal-like in some species.

- *Corolla*: the next whorl toward the apex, composed of units called *petals*, which are typically thin, soft and colored to attract animals that help the process of pollination.

Reproductive

- *Androecium* (from Greek *andros oikia*: man's house): the next whorl (sometimes multiplied into several whorls), consisting of units called stamens. Stamens consist of two parts: a stalk called a filament, topped by an anther where pollen is produced by meiosis and eventually dispersed.

Reproductive parts of Easter Lily (*Lilium longiflorum*). 1. Stigma, 2. Style, 3. Stamens, 4. Filament, 5. Petal

- *Gynoecium* (from Greek *gynaikos oikia*: woman's house): the innermost whorl of a flower, consisting of one or more units called carpels. The carpel or multiple fused carpels form a hollow structure called an ovary, which produces ovules internally. Ovules are megasporangia and they in turn produce megaspores by meiosis which develop into female gametophytes. These give rise to egg cells. The gynoecium of a flower is also described using an alternative terminology wherein the structure one sees in the innermost whorl (consisting of an ovary, style and stigma) is called a pistil. A pistil may consist of a single carpel or a number of carpels fused together. The sticky tip of the pistil, the stigma, is the receptor of pollen. The supportive stalk, the style, becomes the pathway for pollen tubes to grow from pollen grains adhering to the stigma. The relationship to the gynoecium on the receptacle is described as hypogynous (beneath a superior ovary), perigynous (surrounding a superior ovary), or epigynous (above inferior ovary).

Structure

Although the arrangement described above is considered "typical", plant species show a wide variation in floral structure. These modifications have significance in the evolution of flowering plants and are used extensively by botanists to establish relationships among plant species.

The four main parts of a flower are generally defined by their positions on the receptacle and not by their function. Many flowers lack some parts or parts may be modified into other functions and/or look like what is typically another part. In some families, like Ranunculaceae, the petals are greatly reduced and in many species the sepals are colorful and petal-like. Other flowers have modified stamens that are petal-like; the double flowers of Peonies and Roses are mostly petaloid stamens. Flowers show great variation and plant scientists describe this variation in a systematic way to identify and distinguish species.

Specific terminology is used to describe flowers and their parts. Many flower parts are fused together; fused parts originating from the same whorl are connate, while fused parts originating from different whorls are adnate; parts that are not fused are free. When petals are fused into a tube or ring that falls away as a single unit, they are sympetalous (also called gamopetalous).

Connate petals may have distinctive regions: the cylindrical base is the tube, the expanding region is the throat and the flaring outer region is the limb. A sympetalous flower, with bilateral symmetry with an upper and lower lip, is bilabiate. Flowers with connate petals or sepals may have various shaped corolla or calyx, including campanulate, funnelform, tubular, urceolate, salverform or rotate.

Referring to "fusion," as it is commonly done, appears questionable because at least some of the processes involved may be non-fusion processes. For example, the addition of intercalary growth at or below the base of the primordia of floral appendages such as sepals, petals, stamens and carpels may lead to a common base that is not the result of fusion.

Left: A normal zygomorphic Streptocarpus flower. Right: An aberrant peloric Streptocarpus flower.
Both of these flowers appeared on the Streptocarpus hybrid 'Anderson's Crows' Wings'.

Many flowers have a symmetry. When the perianth is bisected through the central axis from any point, symmetrical halves are produced, forming a radial symmetry. These flowers are also known to be actinomorphic or regular, e.g. rose or trillium. When flowers are bisected and produce only one line that produces symmetrical halves the flower is said to be irregular or zygomorphic, e.g. snapdragon or most orchids.

Flowers may be directly attached to the plant at their base (sessile—the supporting stalk or stem is highly reduced or absent). The stem or stalk subtending a flower is called a peduncle. If a peduncle supports more than one flower, the stems connecting each flower to the main axis are called pedicels. The apex of a flowering stem forms a terminal swelling which is called the *torus* or receptacle.

Inflorescence

The familiar calla lily is not a single flower. It is actually an inflorescence of
tiny flowers pressed together on a central stalk that is surrounded by a large petal-like bract.

In those species that have more than one flower on an axis, the collective cluster of flowers is termed an *inflorescence*. Some inflorescences are composed of many small flowers arranged in a formation that resembles a single flower. The common example of this is most members of the very large composite (Asteraceae) group. A single daisy or sunflower, for example, is not a flower but a flower *head*—an inflorescence composed of numerous flowers (or florets). An inflorescence may include specialized stems and modified leaves known as bracts.

Floral Diagrams and Floral Formulae

A *floral formula* is a way to represent the structure of a flower using specific letters, numbers and symbols, presenting substantial information about the flower in a compact form. It can represent a taxon, usually giving ranges of the numbers of different organs, or particular species. Floral formulae have been developed in the early 19th century and their use has declined since. Prenner *et al.* (2010) devised an extension of the existing model to broaden the descriptive capability of the formula. The format of floral formulae differs in different parts of the world, yet they convey the same information.

The structure of a flower can also be expressed by the means of *floral diagrams*. The use of schematic diagrams can replace long descriptions or complicated drawings as a tool for understanding both floral structure and evolution. Such diagrams may show important features of flowers, including the relative positions of the various organs, including the presence of fusion and symmetry, as well as structural details.

Development

A flower develops on a modified shoot or *axis* from a determinate apical meristem (*determinate* meaning the axis grows to a set size). It has compressed internodes, bearing structures that in classical plant morphology are interpreted as highly modified leaves. Detailed developmental studies, however, have shown that stamens are often initiated more or less like modified stems (caulomes) that in some cases may even resemble branchlets. Taking into account the whole diversity in the development of the androecium of flowering plants, we find a continuum between modified leaves (phyllomes), modified stems (caulomes), and modified branchlets (shoots).

Flowering Transition

The transition to flowering is one of the major phase changes that a plant makes during its life cycle. The transition must take place at a time that is favorable for fertilization and the formation of seeds, hence ensuring maximal reproductive success. To meet these needs a plant is able to interpret important endogenous and environmental cues such as changes in levels of plant hormones and seasonable temperature and photoperiod changes. Many perennial and most biennial plants require vernalization to flower. The molecular interpretation of these signals is through the transmission of a complex signal known as florigen, which involves a variety of genes, including constans, flowering locus c and flowering locus T. Florigen is produced in the leaves in reproductively favorable conditions and acts in buds and growing tips to induce a number of different physiological and morphological changes.

The first step of the transition is the transformation of the vegetative stem primordia into floral primordia. This occurs as biochemical changes take place to change cellular differentiation of leaf, bud and stem tissues into tissue that will grow into the reproductive organs. Growth of the central part of the stem tip stops or flattens out and the sides develop protuberances in a whorled or spiral fashion around the outside of the stem end. These protuberances develop into the sepals, petals, stamens, and carpels. Once this process begins, in most plants, it cannot be reversed and the stems develop flowers, even if the initial start of the flower formation event was dependent of some environmental cue. Once the process begins, even if that cue is removed the stem will continue to develop a flower.

Yvonne Aitken has shown that flowering transition depends on a number of factors, and that plants flowering earliest under given conditions had the least dependence on climate whereas later-flowering varieties reacted strongly to the climate setup.

Organ Development

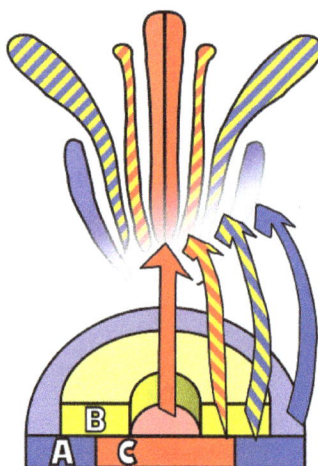

The ABC model of flower development

The molecular control of floral organ identity determination appears to be fairly well understood in some species. In a simple model, three gene activities interact in a combinatorial manner to determine the developmental identities of the organ primordia within the floral meristem. These gene functions are called A, B and C-gene functions. In the first floral whorl only A-genes are expressed, leading to the formation of sepals. In the second whorl both A- and B-genes are expressed, leading to the formation of petals. In the third whorl, B and C genes interact to form stamens and in the center of the flower C-genes alone give rise to carpels. The model is based upon studies of mutants in *Arabidopsis thaliana* and snapdragon, *Antirrhinum majus*. For example, when there is a loss of B-gene function, mutant flowers are produced with sepals in the first whorl as usual, but also in the second whorl instead of the normal petal formation. In the third whorl the lack of B function but presence of C-function mimics the fourth whorl, leading to the formation of carpels also in the third whorl.

Most genes central in this model belong to the MADS-box genes and are transcription factors that regulate the expression of the genes specific for each floral organ.

Floral Function

The principal purpose of a flower is the reproduction of the individual and the species. All flowering plants are *heterosporous*, producing two types of spores. Microspores are produced by meiosis inside anthers while megaspores are produced inside ovules, inside an ovary. In fact, anthers typically consist of four microsporangia and an ovule is an integumented megasporangium. Both types of spores develop into gametophytes inside sporangia. As with all heterosporous plants, the gametophytes also develop inside the spores (are endosporic).

An example of a "perfect flower", this *Crateva religiosa* flower has both stamens (outer ring) and a pistil (center).

In the majority of species, individual flowers have both functional carpels and stamens. Botanists describe these flowers as being *perfect* or *bisexual* and the species as *hermaphroditic*. Some flowers lack one or the other reproductive organ and called *imperfect* or *unisexual*. If unisex flowers are found on the same individual plant but in different locations, the species is said to be *monoecious*. If each type of unisex flower is found only on separate individuals, the plant is *dioecious*.

Flower Specialization and Pollination

Flowering plants usually face selective pressure to optimize the transfer of their pollen, and this is typically reflected in the morphology of the flowers and the behaviour of the plants. Pollen may be transferred between plants via a number of 'vectors'. Some plants make use of abiotic vectors — namely wind (anemophily) or, much less commonly, water (hydrophily). Others use biotic vectors including insects (entomophily), birds (ornithophily), bats (chiropterophily) or other animals. Some plants make use of multiple vectors, but many are highly specialised.

Cleistogamous flowers are self-pollinated, after which they may or may not open. Many Viola and some Salvia species are known to have these types of flowers.

The flowers of plants that make use of biotic pollen vectors commonly have glands called nectaries that act as an incentive for animals to visit the flower. Some flowers have patterns, called nectar guides, that show pollinators where to look for nectar. Flowers also attract pollinators by scent and color. Still other flowers use mimicry to attract pollinators. Some species of orchids, for example, produce flowers resembling female bees in color, shape, and scent. Flowers are also specialized in shape and have an arrangement of the stamens that ensures that pollen grains are transferred to the bodies of the pollinator when it lands in search of its attractant (such as nectar, pollen, or a mate). In pursuing this attractant from many flowers of the same species, the pollinator transfers pollen to the stigmas—arranged with equally pointed precision—of all of the flowers it visits.

Anemophilous flowers use the wind to move pollen from one flower to the next. Examples include grasses, birch trees, ragweed and maples. They have no need to attract pollinators and therefore tend not to be "showy" flowers. Male and female reproductive organs are generally found in separate flowers, the male flowers having a number of long filaments terminating in exposed stamens, and the female flowers having long, feather-like stigmas. Whereas the pollen of animal-pollinated flowers tends to be large-grained, sticky, and rich in protein (another "reward" for pollinators), anemophilous flower pollen is usually small-grained, very light, and of little nutritional value to animals.

Pollination

Grains of pollen sticking to this bee will be transferred to the next flower it visits

The primary purpose of a flower is reproduction. Since the flowers are the reproductive organs of plant, they mediate the joining of the sperm, contained within pollen, to the ovules — contained in the ovary. Pollination is the movement of pollen from the anthers to the stigma. The joining of the sperm to the ovules is called fertilization. Normally pollen is moved from one plant to another, but many plants are able to self pollinate. The fertilized ovules produce seeds that are the next generation. Sexual reproduction produces genetically unique offspring, allowing for adaptation. Flowers have specific designs which encourages the transfer of pollen from one plant to another of the same species. Many plants are dependent upon external factors for pollination, including: wind and animals, and especially insects. Even large animals such as birds, bats, and pygmy possums can be employed. The period of time during which this process can take place (the flower is fully expanded and functional) is called *anthesis*. The study of pollination by insects is called *anthecology*.

Pollen

The types of pollen that most commonly cause allergic reactions are produced by the plain-looking plants (trees, grasses, and weeds) that do not have showy flowers. These plants make small, light, dry pollen grains that are custom-made for wind transport.

The type of allergens in the pollen is the main factor that determines whether the pollen is likely to cause hay fever. For example, pine tree pollen is produced in large amounts by a common tree, which would make it a good candidate for causing allergy. It is, however, a relatively rare cause of allergy because the types of allergens in pine pollen appear to make it less allergenic.

Among North American plants, weeds are the most prolific producers of allergenic pollen. Ragweed is the major culprit, but other important sources are sagebrush, redroot pigweed, lamb's quarters, Russian thistle (tumbleweed), and English plantain.

There is much confusion about the role of flowers in allergies. For example, the showy and entomophilous goldenrod (*Solidago*) is frequently blamed for respiratory allergies, of which it is innocent, since its pollen cannot be airborne. Instead the allergen is usually the pollen of the contemporary bloom of anemophilous ragweed (*Ambrosia*), which can drift for many kilometers.

Scientists have collected samples of ragweed pollen 400 miles out at sea and 2 miles high in the air. A single ragweed plant can generate a million grains of pollen per day.

It is common to hear people say they are allergic to colorful or scented flowers like roses. In fact, only florists, gardeners, and others who have prolonged, close contact with flowers are likely to be sensitive to pollen from these plants. Most people have little contact with the large, heavy, waxy pollen grains of such flowering plants because this type of pollen is not carried by wind but by insects such as butterflies and bees.

Attraction Methods

Plants cannot move from one location to another, thus many flowers have evolved to attract animals to transfer pollen between individuals in dispersed populations. Flowers that are insect-pollinated are called *entomophilous*; literally "insect-loving" in Greek. They can be highly modified along with the pollinating insects by co-evolution. Flowers commonly have glands called *nectaries* on various parts that attract animals looking for nutritious nectar. Birds and bees have color vision, enabling them to seek out "colorful" flowers.

Some flowers have patterns, called nectar guides, that show pollinators where to look for nectar; they may be visible only under ultraviolet light, which is visible to bees and some other insects. Flowers also attract pollinators by scent and some of those scents are pleasant to our sense of smell. Not all flower scents are appealing to humans; a number of flowers are pollinated by insects that are attracted to rotten flesh and have flowers that smell like dead animals, often called Carrion flowers, including *Rafflesia*, the titan arum, and the North American pawpaw (*Asimina triloba*). Flowers pollinated by night visitors, including bats and moths, are likely to concentrate on scent to attract pollinators and most such flowers are white.

Other flowers use mimicry to attract pollinators. Some species of orchids, for example, produce flowers resembling female bees in color, shape, and scent. Male bees move from one such flower to another in search of a mate.

Pollination Mechanism

The pollination mechanism employed by a plant depends on what method of pollination is utilized.

Most flowers can be divided between two broad groups of pollination methods:

Entomophilous: flowers attract and use insects, bats, birds or other animals to transfer pollen from one flower to the next. Often they are specialized in shape and have an arrangement of the stamens that ensures that pollen grains are transferred to the bodies of the pollinator when it lands in search of its attractant (such as nectar, pollen, or a mate). In pursuing this attractant from many flowers of the same species, the pollinator transfers pollen to the stigmas—arranged with equally pointed precision—of all of the flowers it visits. Many flowers rely on simple proximity between flower parts to ensure pollination. Others, such as the *Sarracenia* or lady-slipper orchids, have elaborate designs to ensure pollination while preventing self-pollination.

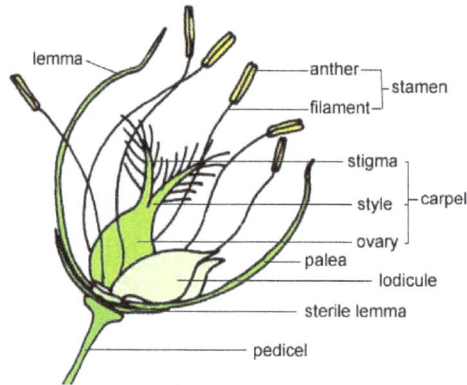

Grass flower with vestigial perianth or lodicules

Anemophilous: flowers use the wind to move pollen from one flower to the next, examples include the grasses, Birch trees, Ragweed and Maples. They have no need to attract pollinators and therefore tend not to grow large blossoms. Whereas the pollen of entomophilous flowers tends to be large-grained, sticky, and rich in protein (another "reward" for pollinators), anemophilous flower pollen is usually small-grained, very light, and of little nutritional value to insects, though it may still be gathered in times of dearth. Honeybees and bumblebees actively gather anemophilous corn (maize) pollen, though it is of little value to them.

Some flowers are self-pollinated and use flowers that never open or are self-pollinated before the flowers open, these flowers are called cleistogamous. Many Viola species and some Salvia have these types of flowers.

Flower-pollinator Relationships

Many flowers have close relationships with one or a few specific pollinating organisms. Many flowers, for example, attract only one specific species of insect, and therefore rely on that insect for successful reproduction. This close relationship is often given as an example of coevolution, as the flower and pollinator are thought to have developed together over a long period of time to match each other's needs.

This close relationship compounds the negative effects of extinction. The extinction of either member in such a relationship would mean almost certain extinction of the other member as well. Some endangered plant species are so because of shrinking pollinator populations.

Fertilization and Dispersal

Some flowers with both stamens and a pistil are capable of self-fertilization, which does increase the chance of producing seeds but limits genetic variation. The extreme case of self-fertilization occurs in flowers that always self-fertilize, such as many dandelions. Conversely, many species of plants have ways of preventing self-fertilization. Unisexual male and female flowers on the same plant may not appear or mature at the same time, or pollen from the same plant may be incapable of fertilizing its ovules. The latter flower types, which have chemical barriers to their own pollen, are referred to as self-sterile or self-incompatible.

Evolution

While land plants have existed for about 425 million years, the first ones reproduced by a simple adaptation of their aquatic counterparts: spores. In the sea, plants—and some animals—can simply scatter out genetic clones of themselves to float away and grow elsewhere. This is how early plants reproduced. But plants soon evolved methods of protecting these copies to deal with drying out and other damage which is even more likely on land than in the sea. The protection became the seed, though it had not yet evolved the flower. Early seed-bearing plants include the ginkgo and conifers. An early fossil of a flowering plant, *Archaefructus liaoningensis* from China, is dated about 125 million years old. Even earlier from China is the 125–130 million years old *Archaefructus sinensis*. Now, another plant (130 million-year-old *Montsechia vidalii*, discovered in Spain) takes the title of world's oldest flower from *Archaefructus sinensis*.

Archaefructus liaoningensis, one of the earliest known flowering plants

Several groups of extinct gymnosperms, particularly seed ferns, have been proposed as the ancestors of flowering plants but there is no continuous fossil evidence showing exactly how flowers evolved. The apparently sudden appearance of relatively modern flowers in the fossil record posed such a problem for the theory of evolution that it was called an "abominable mystery" by Charles Darwin. Recently discovered angiosperm fossils such as *Archaefructus*, along with further discoveries of fossil gymnosperms, suggest how angiosperm characteristics may have been acquired in a series of steps.

Recent DNA analysis (molecular systematics) shows that *Amborella trichopoda*, found on the Pacific island of New Caledonia, is the sister group to the rest of the flowering plants, and morphological studies suggest that it has features which may have been characteristic of the earliest flowering plants.

The general assumption is that the function of flowers, from the start, was to involve animals in the reproduction process. Pollen can be scattered without bright colors and obvious shapes, which would therefore be a liability, using the plant's resources, unless they provide some other benefit. One proposed reason for the sudden, fully developed appearance of flowers is that they evolved in an isolated setting like an island, or chain of islands, where the plants bearing them were able to develop a highly specialized relationship with some specific animal (a wasp, for example), the way many island species develop today. This symbiotic relationship, with a hypothetical wasp bearing pollen from one plant to another much the way fig wasps do today, could have eventually resulted in both the plant(s) and their partners developing a high degree of specialization. Island genetics is believed to be a common source of speciation, especially when it comes to radical adaptations which seem to have required inferior transitional forms.The wasp example is not incidental; bees, apparently evolved specifically for symbiotic plant relationships, are descended from wasps.

Likewise, most fruit used in plant reproduction comes from the enlargement of parts of the flower. This fruit is frequently a tool which depends upon animals wishing to eat it, and thus scattering the seeds it contains.

While many such symbiotic relationships remain too fragile to survive competition with mainland organisms, flowers proved to be an unusually effective means of production, spreading (whatever their actual origin) to become the dominant form of land plant life.

Amborella trichopoda, the sister group to the rest of the flowering plants.

While there is only hard proof of such flowers existing about 130 million years ago, there is some circumstantial evidence that they did exist up to 250 million years ago. A chemical used by plants to defend their flowers, oleanane, has been detected in fossil plants that old, including gigantopterids, which evolved at that time and bear many of the traits of modern, flowering plants, though they are not known to be flowering plants themselves, because only their stems and prickles have been found preserved in detail; one of the earliest examples of petrification.

The similarity in leaf and stem structure can be very important, because flowers are genetically just an adaptation of normal leaf and stem components on plants, a combination of genes normally responsible for forming new shoots. The most primitive flowers are thought to have had a variable number of flower parts, often separate from (but in contact with) each other. The flowers would have tended to grow in a spiral pattern, to be bisexual (in plants, this means both male and female parts on the same flower), and to be dominated by the ovary (female part). As flowers grew more advanced, some variations developed parts fused together, with a much more specific number and design, and with either specific sexes per flower or plant, or at least "ovary inferior".

Flower evolution continues to the present day; modern flowers have been so profoundly influenced by humans that many of them cannot be pollinated in nature. Many modern, domesticated flowers used to be simple weeds, which only sprouted when the ground was disturbed. Some of them tended to grow with human crops, and the prettiest did not get plucked because of their beauty, developing a dependence upon and special adaptation to human affection.

Color

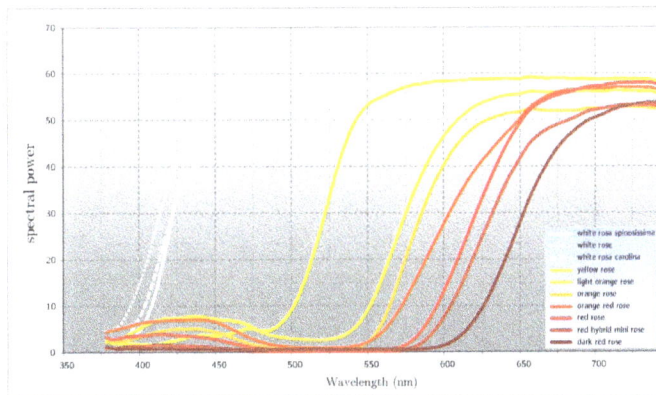

Spectrum of the flowers of rose species. A red rose absorbs about 99.7% of light across a broad area below the red wavelengths of the spectrum, leading to an exceptionally *pure* red. A yellow rose will reflect about 5% of blue light, producing an unsaturated yellow (a yellow with a degree of white in it).

Many flowering plants reflect as much light as possible within the range of visible wavelengths of the pollinator the plant intends to attract. Flowers that reflect the full range of visible light are generally perceived as *white* by a human observer. An important feature of white flowers is that they reflect equally across the visible spectrum. While many flowering plants use white to attract pollinators, the use of color is also widespread (even within the same species). Color allows a flowering plant to be more specific about the pollinator it seeks to attract. The color model used by human color reproduction technology (CMYK) relies on the modulation of pigments that divide the spectrum into broad areas of absorption. Flowering plants by contrast are able to shift the transition point wavelength between absorption and reflection. If it is assumed that the visual systems of most pollinators view the visible spectrum as circular then it may be said that flowering plants produce color by absorbing the light in one region of the spectrum and reflecting the light in the other region. With CMYK, color is produced as a function of the amplitude of the broad regions of absorption. Flowering plants by contrast produce color by modifying the frequency (or rather wavelength) of the light reflected. Most flowers absorb light in the blue to yellow region of the spectrum and reflect light from the green to red region of the spectrum. For many species of flowering plant, it is the transition point that characterizes the color that they produce. Color may be modulated by shifting the transition point between absorption and reflection and in this way a flowering plant may specify which pollinator it seeks to attract. Some flowering plants also have a limited ability to modulate areas of absorption. This is typically not as precise as control over wavelength. Humans observers will perceive this as degrees of saturation (the amount of *white* in the color).

Symbolism

Lilies are often used to denote life or resurrection

Flowers are common subjects of still life paintings, such as this one by Ambrosius Bosschaert the Elder.

Many flowers have important symbolic meanings in Western culture. The practice of assigning meanings to flowers is known as floriography. Some of the more common examples include:

- Red roses are given as a symbol of love, beauty, and passion.

- Poppies are a symbol of consolation in time of death. In the United Kingdom, New Zealand, Australia and Canada, red poppies are worn to commemorate soldiers who have died in times of war.

- Irises/Lily are used in burials as a symbol referring to "resurrection/life". It is also associated with stars (sun) and its petals blooming/shining.

- Daisies are a symbol of innocence.

Flowers within art are also representative of the female genitalia, as seen in the works of artists such as Georgia O'Keeffe, Imogen Cunningham, Veronica Ruiz de Velasco, and Judy Chicago, and in fact in Asian and western classical art. Many cultures around the world have a marked tendency to associate flowers with femininity.

The great variety of delicate and beautiful flowers has inspired the works of numerous poets, especially from the 18th–19th century Romantic era. Famous examples include William Wordsworth's *I Wandered Lonely as a Cloud* and William Blake's *Ah! Sun-Flower*.

Because of their varied and colorful appearance, flowers have long been a favorite subject of visual artists as well. Some of the most celebrated paintings from well-known painters are of flowers, such as Van Gogh's sunflowers series or Monet's water lilies. Flowers are also dried, freeze dried and pressed in order to create permanent, three-dimensional pieces of flower art.

Their symbolism in dreams has also been discussed, with possible interpretations including "blossoming potential".

The Roman goddess of flowers, gardens, and the season of Spring is Flora. The Greek goddess of spring, flowers and nature is Chloris.

In Hindu mythology, flowers have a significant status. Vishnu, one of the three major gods in the Hindu system, is often depicted standing straight on a lotus flower. Apart from the association with Vishnu, the Hindu tradition also considers the lotus to have spiritual significance. For example, it figures in the Hindu stories of creation.

Usage

Flower market – Detroit's Eastern Market

In modern times people have sought ways to cultivate, buy, wear, or otherwise be around flowers and blooming plants, partly because of their agreeable appearance and smell. Around the world, people use flowers for a wide range of events and functions that, cumulatively, encompass one's lifetime:

- For new births or christenings

- As a corsage or boutonniere worn at social functions or for holidays

- As tokens of love or esteem

- For wedding flowers for the bridal party, and for decorations for the hall

- As brightening decorations within the home

- As a gift of remembrance for *bon voyage* parties, welcome-home parties, and "thinking of you" gifts

- For funeral flowers and expressions of sympathy for the grieving

- For worshiping goddesses. In Hindu culture adherents commonly bring flowers as a gift to temples

A woman spreading flowers over a lingam in a temple in Varanasi.

People therefore grow flowers around their homes, dedicate entire parts of their living space to flower gardens, pick wildflowers, or buy flowers from florists who depend on an entire network of commercial growers and shippers to support their trade.

Flowers provide less food than other major plants parts (seeds, fruits, roots, stems and leaves) but they provide several important foods and spices. Flower vegetables include broccoli, cauliflower and artichoke. The most expensive spice, saffron, consists of dried stigmas of a crocus. Other flower spices are cloves and capers. Hops flowers are used to flavor beer. Marigold flowers are fed to chickens to give their egg yolks a golden yellow color, which consumers find more desirable; dried and ground marigold flowers are also used as a spice and colouring agent in Georgian cuisine. Flowers of the dandelion and elder are often made into wine. Bee pollen, pollen collected from bees, is considered a health food by some people. Honey consists of bee-processed flower nectar and is often named for the type of flower, e.g. orange blossom honey, clover honey and tupelo honey.

Hundreds of fresh flowers are edible but few are widely marketed as food. They are often used to add color and flavor to salads. Squash flowers are dipped in breadcrumbs and fried. Edible flowers include nasturtium, chrysanthemum, carnation, cattail, honeysuckle, chicory, cornflower, canna, and sunflower. Some edible flowers are sometimes candied such as daisy, rose, and violet (one may also come across a candied pansy).

Flowers can also be made into herbal teas. Dried flowers such as chrysanthemum, rose, jasmine, camomile are infused into tea both for their fragrance and medical properties. Sometimes, they are also mixed with tea leaves for the added fragrance.

Flowers have been used since as far back as 50,000 years in funeral rituals. Many cultures do draw a connection between flowers and life and death, and because of their seasonal return flowers also suggest rebirth, which may explain why many people place flowers upon graves. In ancient times the Greeks would place a crown of flowers on the head of the deceased as well as cover tombs with wreaths and flower petals. Rich and powerful women in ancient Egypt would wear floral head-dresses and necklaces upon their death as representations of renewal and a joyful afterlife, and the Mexicans to this day use flowers prominently in their Day of the Dead celebrations in the same way that their Aztec ancestors did.

Eight Flowers, a painting by artist Qian Xuan, 13th century, Palace Museum, Beijing.

Sepal

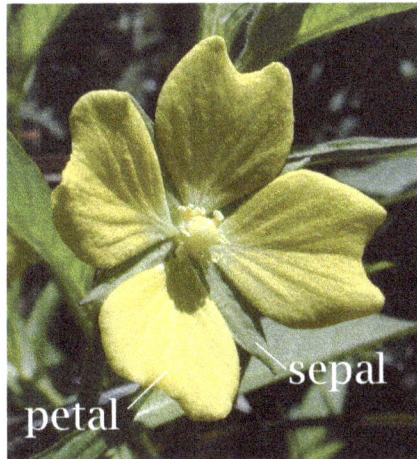

Tetramerous flower of *Ludwigia octovalvis* showing petals and sepals.

After blooming, the sepals of *Hibiscus sabdariffa* expand into an edible accessory fruit.

In many Fabaceae flowers, a calyx tube surrounds the petals.

A sepal is a part of the flower of angiosperms (flowering plants). Usually green, sepals typically function as protection for the flower in bud, and often as support for the petals when in bloom. The term *sepalum* was coined by Noël Martin Joseph de Necker in 1790, and derived from the Greek (*skepi*), a covering.

Collectively the sepals are called the calyx (plural calyces), the outermost whorl of parts that form a flower. The word *calyx* was adopted from the Latin *calyx*, not to be confused with *calix*, a cup or goblet. *Calyx* derived from the Greek (*kalyx*), a bud, a calyx, a husk or wrapping, (cf Sanskrit *kalika*, a bud) while *calix* derived from the Greek (*kylix*), a cup or goblet, and the words have been used interchangeably in botanical Latin.

After flowering, most plants have no more use for the calyx which withers or becomes vestigial. Some plants retain a thorny calyx, either dried or live, as protection for the fruit or seeds. Examples include species of *Acaena*, some of the Solanaceae (for example the Tomatillo, *Physalis philadelphica*), and the water caltrop, *Trapa natans*. In some species the calyx not only persists after flowering, but instead of withering, begins to grow until it forms a bladder-like enclosure around the fruit. This is an effective protection against some kinds of birds and insects, for example in *Hibiscus trionum* and the Cape gooseberry.

Morphologically, both sepals and petals are modified leaves. The calyx (the sepals) and the corolla (the petals) are the outer sterile whorls of the flower, which together form what is known as the *perianth*.

The term *tepal* is usually applied when the parts of the perianth are difficult to distinguish, e.g. the petals and sepals share the same color, or the petals are absent and the sepals are colorful. When the undifferentiated tepals resemble petals, they are referred to as "petaloid", as in petaloid monocots, orders of monocots with brightly coloured tepals. Since they include Liliales, an alternative name is lilioid monocots. Examples of plants in which the term tepal is appropriate include genera such as *Aloe* and *Tulipa*. In contrast, genera such as *Rosa* and *Phaseolus* have well-distinguished sepals and petals.

The number of sepals in a flower is its merosity. Flower merosity is indicative of a plant's classification. The merosity of a eudicot flower is typically four or five. The merosity of a monocot or palaeodicot flower is three, or a multiple of three.

The development and form of the sepals vary considerably among flowering plants. They may be free (polysepalous) or fused together (gamosepalous). Often, the sepals are much reduced, appearing somewhat awn-like, or as scales, teeth, or ridges. Most often such structures protrude until the fruit is mature and falls off.

Examples of flowers with much reduced perianths are found among the grasses.

In some flowers, the sepals are fused towards the base, forming a calyx tube (as in the Lythraceae family, and Fabaceae). In other flowers (e.g., Rosaceae, Myrtaceae) a hypanthium includes the bases of sepals, petals, and the attachment points of the stamens.

Petal

Petals are modified leaves that surround the reproductive parts of flowers. They are often brightly colored or unusually shaped to attract pollinators. Together, all of the petals of a flower are called a corolla. Petals are usually accompanied by another set of special leaves called sepals, that collectively form the *calyx* and lie just beneath the corolla. The calyx and the corolla together make up the perianth. When the petals and sepals of a flower are difficult to distinguish, they are collectively called tepals. Examples of plants in which the term *tepal* is appropriate include genera such as *Aloe* and *Tulipa*. Conversely, genera such as *Rosa* and *Phaseolus* have well-distinguished sepals and petals. When the undifferentiated tepals resemble petals, they are referred to as "petaloid", as in petaloid monocots, orders of monocots with brightly coloured tepals. Since they include Liliales, an alternative name is lilioid monocots.

Although petals are usually the most conspicuous parts of animal-pollinated flowers, wind-pollinated species, such as the grasses, either have very small petals or lack them entirely.

A Tulip's actinomorphic flower with three petals and three sepals, that collectively present a good example of an undifferentiated perianth. In this case, the word "tepals" is used.

Corolla

Corolla forming a tube with long points (and a separate green calyx tube).

The role of the corolla in plant evolution has been studied extensively since Charles Darwin postulated a theory of the origin of elongated corollae and corolla tubes.

If the petals are free from one another in the corolla, the plant is *polypetalous* or *choripetalous*; while if the petals are at least partially fused together, it is *gamopetalous* or *sympetalous*. The corolla in some plants forms a tube.

Variations

Pelargonium peltatum: its floral structure is almost identical to that of geraniums, but it is conspicuously zygomorphic

Petals can differ dramatically in different species. The number of petals in a flower may hold clues to a plant's classification. For example, flowers on eudicots (the largest group of dicots) most frequently have four or five petals while flowers on monocots have three or six petals, although there are many exceptions to this rule.

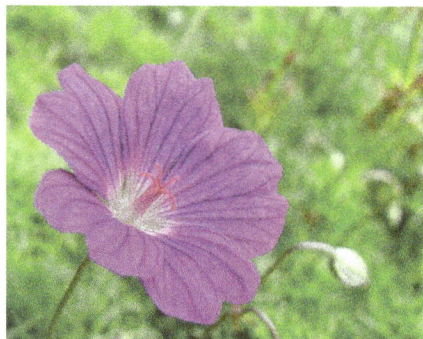

Geranium incanum, with an actinomorphic flower typical of the genus.

White pea, a zygomorphic flower

Narcissus pseudonarcissus, showing from the bend to the tip of the flower:
spathe, floral cup, tepals, corona.

The petal whorl or corolla may be either radially or bilaterally symmetrical. If all of the petals are essentially identical in size and shape, the flower is said to be regular or actinomorphic (meaning "ray-formed"). Many flowers are symmetrical in only one plane (i.e., symmetry is bilateral) and are termed irregular or zygomorphic (meaning "yoke-" or "pair-formed"). In *irregular* flowers, other floral parts may be modified from the *regular* form, but the petals show the greatest deviation from radial symmetry. Examples of zygomorphic flowers may be seen in orchids and members of the pea family.

In many plants of the aster family such as the sunflower, *Helianthus annuus*, the circumference of the flower head is composed of ray florets. Each ray floret is anatomically an individual flower with a single large petal. Florets in the centre of the disc typically have no or very reduced petals. In some plants such as *Narcissus* the lower part of the petals or tepals are fused to form a floral cup (hypanthium) above the ovary, and from which the petals proper extend.

Petal often consists of two parts: the upper, broad part, similar to leaf blade, called the *limb* and the lower part, narrow, similar to leaf petiole, called the *claw*. Claws are developed in petals of some flowers of the family *Brassicaceae*, such as *Erysimum cheiri*.

The inception and further development of petals shows a great variety of patterns. Petals of different species of plants vary greatly in colour or colour pattern, both in visible light and in ultraviolet. Such patterns often function as guides to pollinators, and are variously known as nectar guides, pollen guides, and floral guides.

Genetics

The genetics behind the formation of petals, in accordance with the ABC model of flower development, are that sepals, petals, stamens, and carpels are modified versions of each other. It appears that the mechanisms to form petals evolved very few times (perhaps only once), rather than evolving repeatedly from stamens.

Significance of Pollination

Pollination is an important step in the sexual reproduction of higher plants. Pollen is produced by the male flower or by the male organs of hermaphroditic flowers.

Pollen does not move on its own and thus requires wind or animal pollinators to disperse the pollen to the stigma (botany) of the same or nearby flowers. However, pollinators are rather selective in determining the flowers they choose to pollinate. This develops competition between flowers and as a result flowers must provide incentives to appeal to pollinators (unless the flower self-pollinates or is involved in wind pollination). Petals play a major role in competing to attract pollinators. Henceforth pollination dispersal could occur and the survival of many species of flowers could prolong.

Functions and Purposes of Petals

Petals have various functions and purposes depending on the type of plant. In general, petals operate to protect some parts of the flower and attract/repel specific pollinators. Flower Petal Function: This is where the positioning of the flower petals are located on the flower is the corolla e.g. the buttercup having shiny yellow flower petals which contain guidelines amongst the petals in aiding the pollinator towards the nectar. Pollinators have the ability to determine specific flowers they wish to pollinate. Using incentives flowers draw pollinators and set up a mutual relation between each other in which case the pollinators will remember to always guard and pollinate these flowers (unless incentives are not consistently met and competition prevails).

Scent

The petals could produce different scents to allure desirable pollinators and/or repel undesirable pollinators. Some flowers will also mimic the scents produced by materials such as decaying meat, to attract pollinators to them.

Colour

Various colour traits are used by different petals that could attract pollinators that have poor smelling abilities, or that only come out at certain parts of the day. Some flowers are able to change the colour of their petals as a signal to mutual pollinators to approach or keep away.

Shape and Size

Furthermore, the shape and size of the flower/petals is important in selecting the type of pollinators they need. For example, large petals and flowers will attract pollinators at a large distance and/or that are large themselves. Collectively the scent, colour and shape of petals all play a role in attracting/repelling specific pollinators and providing suitable conditions for pollinating. Some pollinators include insects, birds, bats and the wind. In some petals, a distinction can be made between a lower narrowed, stalk-like basal part referred to as the claw, and a wider distal part referred to as the blade. Often the claw and blade are at an angle with one another.

Types of Pollination

Wind Pollination

Wind-pollinated flowers often have small dull petals and produce little or no scent. Some of these flowers will often have no petals at all. Flowers that depend on wind pollination will produce large amounts of pollen because most of the pollen scattered by the wind tends to not reach other flowers.

Attracting Insects

Flowers have various regulatory mechanisms in order to attract insects. One such helpful mechanism is the use of colour guiding marks. Insects such as the bee or butterfly can see the ultraviolet marks which are contained on these flowers, acting as an attractive mechanism which is not visible towards the human eye. Many flowers contain a variety of shapes acting to aid with the landing of the visiting insect and also influence the insect to brush against anthers and stigmas (parts of the flower). One such example of a flower is the pōhutukawa (*Metrosideros excelsa*) which acts to regulate colour within a different way. The pōhutukawa contains small petals also having bright large red clusters of stamens. Another attractive mechanism for flowers is the use of scent which is highly attractive towards humans such as the rose, but some are very fragrant within attracting flies as they have a smell of rotting meat. Dark is another factor in which flowers have grown to adapt these conditions so colour lacks vision at night therefore scent is the solution for flowers which are pollinated by night flying insects such as the moth.

Attracting Birds

Flowers are also pollinated by birds and must be large and colorful to be visible against natural scenery. Such bird –pollinated native plants include: Kōwhai (*Sophora* species), flax (*Phormium tenax*, harakeke) and kākā beak (*Clianthus puniceus*, kōwhai ngutu-kākā). Interestingly enough, flowers adapt the mechanism on their petals to change colour in acting as a communicative mechanism for the bird to visit. An example is the tree fuchsia (*Fuchsia excorticata*, kōtukutuku) which are green when needing to be pollinated and turn red for the birds to stop coming and pollinating the flower.

Bat-pollinated Flowers

Flowers can be pollinated by short tailed bats. An example of this is the dactylanthus (*Dactylanthus taylorii*). This plant has its home under the ground acting the role of a parasite on the roots of forest trees. The dactylanthus has only its flowers pointing to the surface and the flowers lack colour but have the advantage of containing lots of nectar and a very strong scent. These act as a very useful mechanism in attracting the bat.

Stamen

Stamens of a *Hippeastrum* with white filaments and prominent anthers carrying pollen.

The stamen (plural *stamina* or *stamens*) is the pollen-producing reproductive organ of a flower. Collectively the stamens form the androecium.

Morphology and Terminology

A stamen typically consists of a stalk called the filament and an anther which contains *microsporangia*. Most commonly anthers are two-lobed and are attached to the filament either at the base or in the middle area of the anther. The sterile tissue between the lobes is called the connective. A pollen grain develops from a microspore in the microsporangium and contains the male gametophyte.

The stamens in a flower are collectively called the *androecium*. The androecium can consist of as few a one-half stamen (i.e. a single locule) as in *Canna* species or as many as 3,482 stamens which have been counted in *Carnegiea gigantea*. The androecium in various species of plants forms a great variety of patterns, some of them highly complex. It surrounds the gynoecium and is surrounded by the perianth. A few members of the family Triuridaceae, particularly *Lacandonia schismatica*, are exceptional in that their gynoecia surround their androecia.

Hippeastrum flowers showing stamens above the style (with its terminal stigma).

Etymology

- Stamen is the Latin word meaning "thread" (originally thread of the warp, in weaving).

- Filament derives from classical Latin *filum*, meaning "thread"

- Anther derives from French *anthère*, from classical Latin *anthera*, meaning "medicine extracted from the flower" in turn from Ancient Greek feminine of «flowery», from "flower"

- Androecium derives from Ancient Greek meaning man , and house or chamber/room.

Variation in Morphology

Depending on the species of plant, some or all of the stamens in a flower may be attached to the petals or to the floral axis. They also may be free-standing or fused to one another in many different ways, including fusion of some but not all stamens. The filaments may be fused and the anthers free, or the filaments free and the anthers fused. Rather than there being two locules, one locule of a stamen may fail to develop, or alternatively the two locules may merge late in development to give a single locule. Extreme cases of stamen fusion occur in some species of *Cyclanthera* in the family Cucurbitaceae and in section *Cyclanthera* of genus *Phyllanthus* (family Euphorbiaceae) where the stamens form a ring around the gynoecium, with a single locule.

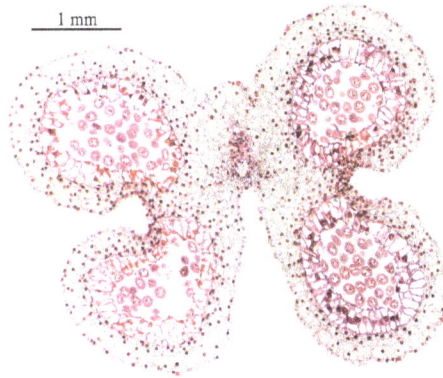

Cross section of a *Lilium* stamen, with four locules surrounded by the tapetum.

Pollen Production

A typical anther contains four microsporangia. The *microsporangia* form sacs or pockets (*locules*) in the anther (anther sacs or pollen sacs). The two separate locules on each side of an anther may fuse into a single locule. Each microsporangium is lined with a nutritive tissue layer called the *tapetum* and initially contains diploid pollen mother cells. These undergo meiosis to form haploid spores. The spores may remain attached to each other in a tetrad or separate after meiosis. Each microspore then divides mitotically to form an immature microgametophyte called a pollen grain.

The pollen is eventually released when the anther forms openings (dehisces). These may consist of longitudinal slits, pores, as in the heath family (Ericaceae), or by valves, as in the barberry family (Berberidaceae). In some plants, notably members of Orchidaceae and Asclepiadoideae, the pollen remains in masses called pollinia, which are adapted to attach to particular pollinating agents such as birds or insects. More commonly, mature pollen grains separate and are dispensed by wind or water, pollinating insects, birds or other pollination vectors.

Pollen of angiosperms must be transported to the stigma, the receptive surface of the *carpel*, of a compatible flower, for successful pollination to occur. After arriving, the pollen grain (an immature microgametophyte) typically completes its development. It may grow a pollen tube and undergoing mitosis to produce two sperm nuclei.

Sexual Reproduction in Plants

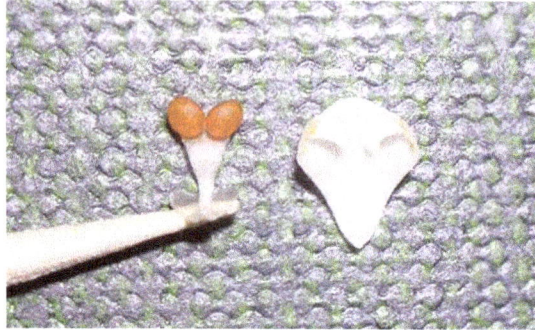

Stamen with pollinia and its anther cap. *Phalaenopsis* orchid.

In the typical flower (that is, in the majority of flowering plant species) each flower has both carpels and stamens. In some species, however, the flowers are unisexual with only carpels or stamens. (monoecious = both types of flowers found on the same plant; dioecious = the two types of flower found only on different plants). A flower with only stamens is called androecious. A flower with only carpels is called gynoecious.

A flower having only functional stamens and lacking functional carpels is called a staminate flower, or (inaccurately) male. A plant with only functional carpels is called pistillate, or (inaccurately) female.

An abortive or rudimentary stamen is called a staminodium or staminode, such as in *Scrophularia nodosa*.

The carpels and stamens of orchids are fused into a column. The top part of the column is formed by the anther, which is covered by an anther cap.

Descriptive Terms

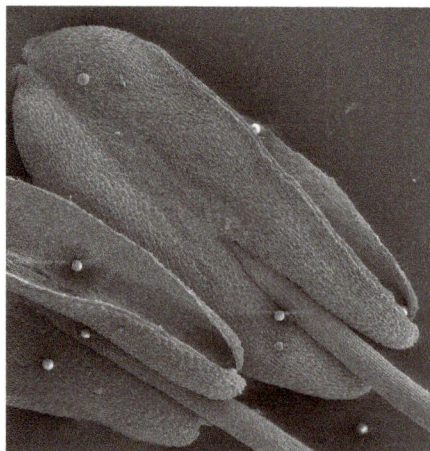

Scanning electron microscope image of *Pentas lanceolata* anthers, with pollen grains on surface.

Lily stamens with prominent red anthers and white filaments.

- A column formed from the fusion of multiple filaments is known as an androphore.

The anther can be attached to the filament's connective in two ways:

- basifixed: attached at its base to the filament

- pseudobasifixed: a somewhat misnomer configuration where connective tissue extends in a tube around the filament

- dorsifixed: attached at its center to the filament, usually versatile (able to move)

Stamens can be connate (fused or joined in the same whorl):

- extrorse: anther dehiscence directed away from the centre of the flower. Cf. introrse, directed inwards, and latrorse towards the side.

- monadelphous: fused into a single, compound structure

- declinate: curving downwards, then up at the tip (also - declinate-descending)

- diadelphous: joined partially into two androecial structures

- pentadelphous: joined partially into five androecial structures

- synandrous: only the anthers are connate (such as in the Asteraceae). The fused stamens are referred to as a synandrium.

Stamens can also be adnate (fused or joined from more than one whorl):

- epipetalous: adnate to the corolla

- epiphyllous: adnate to undifferentiated tepals (as in many Liliaceae)

They can have different lengths from each other:

- didymous: two equal pairs

- didynamous: occurring in two pairs, a long pair and a shorter pair

- tetradynamous: occurring as a set of six stamens with four long and two shorter ones

or respective to the rest of the flower (perianth):

- exserted: extending beyond the corolla

- included: not extending beyond the corolla

They may be arranged in one of two different patterns:

- spiral; or

- whorled: one or more discrete whorls (series)

They may be arranged, with respect to the petals:

- diplostemonous: in two whorls, the outer alternating with the petals, while the inner is opposite the petals.

- obdiplostemonous: in two whorls, the outer opposite the petals

Gynoecium

Gynoecium (from Ancient Greek *gyne*, meaning *woman*, and meaning *house*) is most commonly used as a collective term for the parts of a flower that produce ovules and ultimately develop into the fruit and seeds. The gynoecium is the innermost whorl of (one or more) pistils in a flower and is typically surrounded by the pollen-producing reproductive organs, the stamens, collectively called the androecium. The gynoecium is often referred to as the "female" portion of the flower, although rather than directly producing female gametes (i.e. egg cells), the gynoecium produces megaspores, each of which develops into a female gametophyte which then produces egg cells.

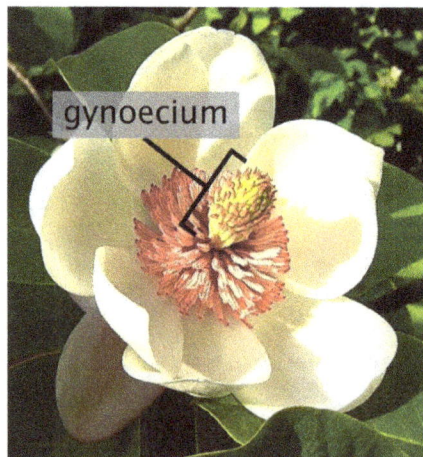

Flower of *Magnolia × wieseneri* showing the many pistils making up the gynoecium in the middle of the flower.

The term gynoecium is also used by botanists to refer to a cluster of archegonia and any associated modified leaves or stems present on a gametophyte shoot in mosses, liverworts and hornworts.

The corresponding terms for the male parts of those plants are clusters of antheridia within the androecium.

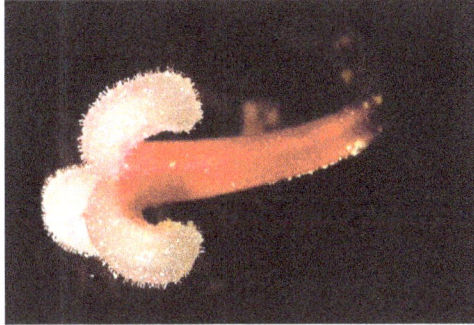
Hippeastrum stigmas and style

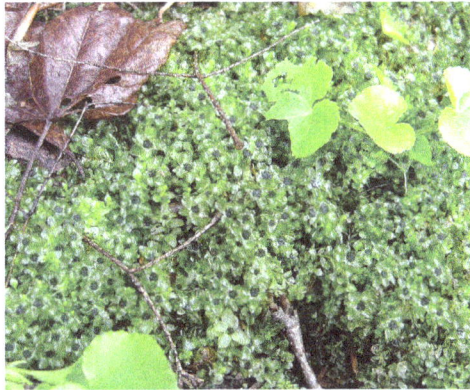
Moss plants with gynoecia, clusters of archegonia at the apex of each shoot.

Flowers that bear a gynoecium but no stamens are called carpellate. Flowers lacking a gynoecium are called staminate.

The gynoecium is often referred to as female because it gives rise to female (egg-producing) gametophytes, however, strictly speaking sporophytes do not have a sex, only gametophytes do.

Gynoecium development and arrangement is important in systematic research and identification of angiosperms, but can be the most challenging of the floral parts to interpret.

Pistils

The gynoecium may consist of one or more separate pistils. A pistil typically consists of an expanded basal portion called the ovary, an elongated section called a style and an apical structure that receives pollen called a stigma.

- The ovary (from Latin *ovum* meaning egg), is the enlarged basal portion which contains placentas, ridges of tissue bearing one or more ovules (integumented megasporangia). The placentas and/or ovule(s) may be born on the gynoecial appendages or less frequently on the floral apex. The chamber in which the ovules develop is called a locule (or sometimes cell).

- The style (from Ancient Greek stülos meaning a pillar), is a pillar-like stalk through which pollen tubes grow to reach the ovary. Some flowers such as *Tulipa* do not have a distinct style, and the stigma sits directly on the ovary. The style is a hollow tube in some plants such as lilies, or has transmitting tissue through which the pollen tubes grow.

- The stigma (from Ancient Greek *stigma* meaning mark, or puncture), is usually found at the tip of the style, the portion of the carpel(s) that receives pollen (male gametophytes). It is commonly sticky or feathery to capture pollen.

The word "pistil" comes from Latin *pistillum* meaning pestle. A sterile pistil in a male flower is referred to as a pistillode.

Carpels

The pistils of a flower are considered to be composed of carpels. A carpel is a theoretical construct interpreted as modified leaves bearing structures called ovules, inside which the egg cells ultimately form. A pistil may consist of one carpel, with its ovary, style and stigma, or several carpels may be joined together with a single ovary, the whole unit called a pistil. The gynoecium may consist of one or more uni-carpellate (with one carpel) pistils, or of one multi-carpellate pistil. The number of carpels is described by terms such as tricarpellate (three carpels).

Carpels are thought to be phylogenetically derived from ovule-bearing leaves or leaf homologues (megasporophylls), which evolved to form a closed structure containing the ovules. This structure is typically rolled and fused along the margin.

Although many flowers satisfy the above definition of a carpel, there are also flowers that do not have carpels according to this definition because in these flowers the ovule(s), although enclosed, are borne directly on the shoot apex, and only later become enclosed by the carpel. Different remedies have been suggested for this problem. An easy remedy that applies to most cases is to redefine the carpel as an appendage that encloses ovule(s) and may or may not bear them.

Centre of a *Ranunculus repens* (Creeping Buttercup) showing multiple unfused carpels surrounded by longer stamens.

Cross-section through the ovary of *Narcissus* showing multiple connate carpels (a compound pistil) fused along the placental line where the ovules form in each locule.

Types of Gynoecia

If a gynoecium has a single carpel, it is called monocarpous. If a gynoecium has multiple, distinct (free, unfused) carpels, it is apocarpous. If a gynoecium has multiple carpels "fused" into a single structure, it is syncarpous. A syncarpous gynoecium can sometimes appear very much like a monocarpous gynoecium.

The degree of connation ("fusion") in a syncarpous gynoecium can vary. The carpels may be "fused" only at their bases, but retain separate styles and stigmas. The carpels may be "fused" entirely, except for retaining separate stigmas. Sometimes (e.g., Apocynaceae) carpels are fused by their styles or stigmas but possess distinct ovaries. In a syncarpous gynoecium, the "fused" ovaries of the constituent carpels may be referred to collectively as a single compound ovary. It can be a challenge to determine how many carpels fused to form a syncarpous gynoecium. If the styles and stigmas are distinct, they can usually be counted to determine the number of carpels. Within the compound ovary, the carpels may have distinct locules divided by walls called septa. If a syncarpous gynoecium has a single style and stigma and a single locule in the ovary, it may be necessary to examine how the ovules are attached. Each carpel will usually have a distinct line of placentation where the ovules are attached.

Pistil Development

Pistils begin as small primordia on a floral apical meristem, forming later than, and closer to the (floral) apex than sepal, petal and stamen primordia. Morphological and molecular studies of pistil ontogeny reveal that carpels are most likely homologous to leaves.

A carpel has a similar function to a megasporophyll, but typically includes a stigma, and is fused, with ovules enclosed in the enlarged lower portion, the ovary.

In some basal angiosperm lineages, Degeneriaceae and Winteraceae, a carpel begins as a shallow cup where the ovules develop with laminar placentation, on the upper surface of the carpel. The carpel eventually forms a folded, leaf-like structure, not fully sealed at its margins. No style exists, but a broad stigmatic crest along the margin allows pollen tubes access along the surface and between hairs at the margins.

Two kinds of fusion have been distinguished: postgenital fusion that can be observed during the development of flowers, and congenital fusion that cannot be observed i.e., fusions that occurred during phylogeny. But it is very difficult to distinguish fusion and non-fusion processes in the evolution of flowering plants. Some processes that have been considered congenital (phylogenetic) fusions appear to be non-fusion processes such as, for example, the de novo formation of intercalary growth in a ring zone at or below the base of primordia. Therefore, "it is now increasingly acknowledged that the term 'fusion,' as applied to phylogeny (as in 'congenital fusion') is ill-advised."

Gynoecium Position

Basal angiosperm groups tend to have carpels arranged spirally around a conical or dome-shaped receptacle. In later lineages, carpels tend to be in whorls.

The relationship of the other flower parts to the gynoecium can be an important systematic and taxonomic character. In some flowers, the stamens, petals, and sepals are often said to be "fused" into a "floral tube" or hypanthium. However, as Leins & Erbar (2010) pointed out, "the classical view that the wall of the inferior ovary results from the "congenital" fusion of dorsal carpel flanks and the floral axis does not correspond to the ontogenetic processes that can actually be observed. All that can be seen is an intercalary growth in a broad circular zone that changes the shape of the floral axis (receptacle)." And what happened during evolution is not a phylogenetic fusion but the formation of a unitary intercalary meristem. Evolutionary developmental biology investigates such developmental processes that arise or change during evolution.

Flowers and fruit (capsules) of the ground orchid, *Spathoglottis plicata*, illustrating an inferior ovary.

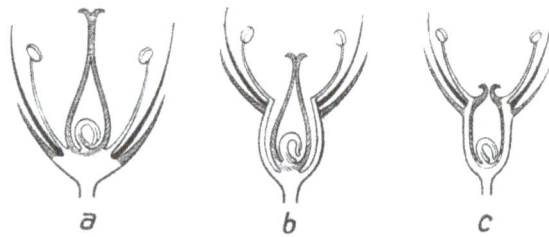

Fig. 239. Schematische Darstellung einer hypogynen (*a*), perigynen (*b*) und epigynen (*c*) Blüte.

Illustration showing longitudinal sections through hypogynous (a), perigynous (b), and epigynous (c) flowers.

If the hypanthium is absent, the flower is *hypogynous*, and the stamens, petals, and sepals are all attached to the receptacle below the gynoecium. Hypogynous flowers are often referred to as having a *superior ovary*. This is the typical arrangement in most flowers.

If the hypanthium is present up to the base of the style(s), the flower is *epigynous*. In an epigynous flower, the stamens, petals, and sepals are attached to the hypanthium at the top of the ovary or, occasionally, the hypanthium may extend beyond the top of the ovary. Epigynous flowers are often referred to as having an *inferior ovary*. Plant families with epigynous flowers include orchids, asters, and evening primroses.

Between these two extremes are *perigynous* flowers, in which a hypanthium is present, but is either free from the gynoecium (in which case it may appear to be a cup or tube surrounding the gynoecium) or connected partly to the gynoecium (with the stamens, petals, and sepals attached to the hypanthium part of the way up the ovary). Perigynous flowers are often referred to as having a *half-inferior ovary* (or, sometimes, *partially inferior* or *half-superior*). This arrangement is particularly frequent in the rose family and saxifrages.

Occasionally, the gynoecium is born on a stalk, called the gynophore, as in *Isomeris arborea*.

Placentation

Within the ovary, each ovule is born by a placenta or arises as a continuation of the floral apex. The placentas often occur in distinct lines called lines of placentation. In monocarpous or apocarpous gynoecia, there is typically a single line of placentation in each ovary. In syncarpous gynoecia, the lines of placentation can be regularly spaced along the wall of the ovary (parietal placentation), or near the center of the ovary. In the latter case, separate terms are used depending on whether or not the ovary is divided into separate locules. If the ovary is divided, with the ovules born on a line of placentation at the inner angle of each locule, this is axile placentation. An ovary with free central placentation, on the other hand, consists of a single compartment without septae and the ovules are attached to a central column that arises directly from the floral apex (axis). In some cases a single ovule is attached to the bottom or top of the locule (basal or apical placentation, respectively).

The Ovule

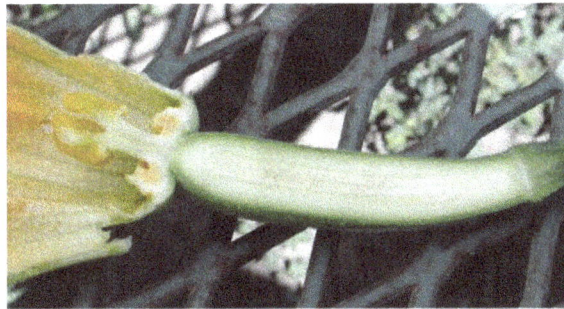

Longitudinal section of carpellate flower of squash showing ovary, ovules, stigma, style, and petals.

In flowering plants, the *ovule* (from Latin *ovulum* meaning small egg) is a complex structure born inside ovaries. The ovule initially consists of a stalked, integumented *megasporangium* (also called the *nucellus*). Typically, one cell in the megasporangium undergoes meiosis resulting in one to four megaspores. These develop into a megagametophyte (often called the embryo sac) within the ovule. The megagametophyte typically develops a small number of cells, including two special cells, an egg cell and a binucleate central cell, which are the gametes involved in double fertilization. The central cell, once fertilized by a sperm cell from the pollen becomes the first cell of the endosperm, and the egg cell once fertilized become the zygote that develops into the embryo. The gap in the integuments through which the pollen tube enters to deliver sperm to the egg is called the micropyle. The stalk attaching the ovule to the placenta is called the funiculus.

Role of the Stigma and Style

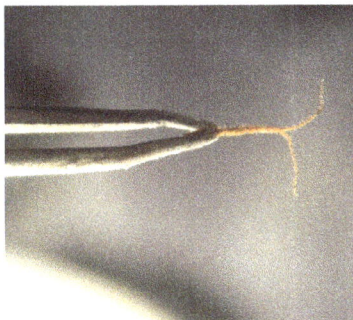

Stigmas and style of *Cannabis sativa* held in a pair of forceps.

Stigma of a *Crocus* flower.

Stigmas can vary from long and slender to globe-shaped to feathery. The stigma is the receptive tip of the carpel(s), which receives pollen at pollination and on which the pollen grain germinates. The stigma is adapted to catch and trap pollen, either by combining pollen of visiting insects or by various hairs, flaps, or sculpturings.

The style and stigma of the flower are involved in most types of self incompatibility reactions. Self-incompatibility, if present, prevents fertilization by pollen from the same plant or from genetically similar plants, and ensures outcrossing.

Leaf

Leaf of *Tilia tomentosa* (Silver lime tree)

A leaf is an organ of a vascular plant and is the principal lateral appendage of the stem. The leaves and stem together form the shoot. Leaves are collectively referred to as foliage, as in "autumn foliage".

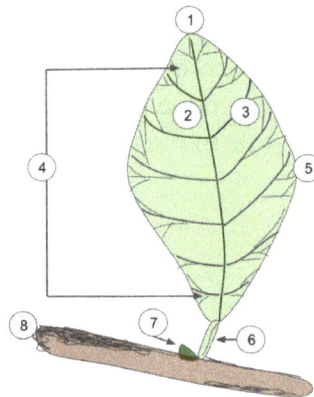

Diagram of a simple leaf

1. Apex
2. Midvein (Primary vein)
3. Secondary vein.
4. Lamina.
5. Leaf margin
6. Petiole
7. Bud
8. Stem

Although leaves can be seen in many different shapes, sizes and textures, typically a leaf is a thin, dorsiventrally flattened organ, borne above ground and specialized for photosynthesis. In most leaves, the primary photosynthetic tissue, the palisade mesophyll, is located on the upper side of the blade or lamina of the leaf but in some species, including the mature foliage of *Eucalyptus*, palisade mesophyll is present on both sides and the leaves are said to be isobilateral. Most leaves have distinctive upper surface (*adaxial*) and lower surface (*abaxial*) that differ in colour, hairiness, the number of stomata (pores that intake and output gases), epicuticular wax amount and structure and other features.

Broad, flat leaves with complex venation are known as *megaphylls* and the species that bear them, the majority, as broad-leaved or megaphyllous plants. In others, such as the clubmosses, with different evolutionary origins, the leaves are simple, with only a single vein and are known as microphylls.

Some leaves, such as bulb scales are not above ground, and in many aquatic species the leaves are submerged in water. Succulent plants often have thick juicy leaves, but some leaves are without major photosynthetic function and may be dead at maturity, as in some cataphylls and spines. Furthermore, several kinds of leaf-like structures found in vascular plants are not totally homologous with them. Examples include flattened plant stems called phylloclades and cladodes, and flattened leaf stems called phyllodes which differ from leaves both in their structure and origin. Many structures of non-vascular plants, such as the phyllids of mosses and liverworts and even of some foliose lichens, which are not plants at all (in the sense of being members of the kingdom Plantae), look and function much like leaves.

General Characteristics

Leaves are the most important organs of most vascular plants. Since plants are autotrophic, they do not need food from other living things to survive but instead use carbon dioxide, water and light energy, to create their own organic matter by photosynthesis of simple sugars, such as glucose and sucrose. These are then further processed by chemical synthesis into more complex organic molecules such as cellulose, the basic structural material in plant cell walls. The plant must therefore bring these three ingredients together in the leaf for photosynthesis to take place. The leaves draw water from the ground in the transpiration stream through a vascular conducting system known as xylem and obtain carbon dioxide from the atmosphere by diffusion through openings called stomata in the outer covering layer of the leaf (epidermis), while leaves are orientated to maximise their exposure to sunlight. Once sugar has been synthesized, it needs to be transported to areas of active growth such as the plant shoots and roots. Vascular plants transport sucrose in a special tissue called the phloem. The phloem and xylem are parallel to each other but the transport of materials is usually in opposite directions. Within the leaf these vascular systems branch (ramify) to form veins which supply as much as the leaf as possible, ensuring that cells carrying out photosynthesis are close to the transportation system.

Typically leaves are broad, flat and thin (dorsiventrally flattened), thereby maximising the surface area directly exposed to light and enabling the light to penetrate the tissues and reach the chloroplasts, thus promoting photosynthesis. They are arranged on the plant so as to expose their surfaces to light as efficiently as possible without shading each other, but there are many exceptions and complications. For instance plants adapted to windy conditions may have pendent leaves, such as in many willows and eucalyptss. The flat, or laminar, shape also maximises thermal contact with the surrounding air, promoting cooling. Functionally, in addition to photosynthesis the leaf is the principal site of transpiration and guttation.

Many gymnosperms have thin needle-like or scale-like leaves that can be advantageous in cold climates with frequent snow and frost. These are interpreted as reduced from megaphyllous leaves of their Devonian ancestors. Some leaf forms are adapted to modulate the amount of light they absorb to avoid or mitigate excessive heat, ultraviolet damage, or desiccation, or to sacrifice light-absorption efficiency in favour of protection from herbivory. For xerophytes the major constraint is not light flux or intensity, but drought. Some window plants such as *Fenestraria* species and some *Haworthia* species such as *Haworthia tesselata* and *Haworthia truncata* are examples of xerophytes. and *Bulbine mesembryanthemoides*.

Leaves also function to store chemical energy and water (especially in succulents) and may become specialised organs serving other functions, such as tendrils of peas and other legumes, the protective spines of cacti and the insect traps in carnivorous plants such as *Nepenthes* and *Sarracenia*. Leaves are the fundamental structural units from which cones are constructed in gymnosperms (each cone scale is a modified megaphyll leaf known as a sporophyll) and from which flowers are constructed in flowering plants.

The internal organisation of most kinds of leaves has evolved to maximise exposure of the photosynthetic organelles, the chloroplasts, to light and to increase the absorption of carbon dioxide while at the same time controlling water loss. Their surfaces are waterproofed by the plant cuticle and gas exchange between the mesophyll cells and the atmosphere is controlled by minute openings called stomata, about 10 μm which open or close to regulate the rate exchange of carbon dioxide, oxygen, and water vapour into and out of the internal intercellular space system. Stomatal opening is controlled by the turgor pressure in a pair of guard cells that surround the stomatal aperture. In any square centimeter of a plant leaf there may be from 1,000 to 100,000 stomata.

Vein skeleton of a leaf. Veins contain lignin that make them harder to degrade for microorganisms.

The shape and structure of leaves vary considerably from species to species of plant, depending largely on their adaptation to climate and available light, but also to other factors such as grazing animals (such as deer), available nutrients, and ecological competition from other plants. Considerable changes in leaf type occur within species too, for example as a plant matures; as a case in point *Eucalyptus* species commonly have isobilateral, pendent leaves when mature and dominating their neighbours; however, such trees tend to have erect or horizontal dorsiventral leaves as seedlings, when their growth is limited by the available light. Other factors include the need to balance water loss at high temperature and low humidity against the need to absorb atmospheric carbon dioxide. In most plants leaves also are the primary organs responsible for transpiration and guttation (beads of fluid forming at leaf margins).

Near the ground these *Eucalyptus* saplings have juvenile dorsiventral foliage from the previous year, but this season their newly sprouting foliage is isobilateral, like the mature foliage on the adult trees above.

Leaves can also store food and water, and are modified accordingly to meet these functions, for example in the leaves of succulent plants and in bulb scales. The concentration of photosynthetic structures in leaves requires that they be richer in protein, minerals, and sugars than, say, woody stem tissues. Accordingly, leaves are prominent in the diet of many animals.

Correspondingly, leaves represent heavy investment on the part of the plants bearing them, and their retention or disposition are the subject of elaborate strategies for dealing with pest pressures, seasonal conditions, and protective measures such as the growth of thorns and the production of phytoliths, lignins, tannins and poisons.

Deciduous plants in frigid or cold temperate regions typically shed their leaves in autumn, whereas in areas with a severe dry season, some plants may shed their leaves until the dry season ends. In either case the shed leaves may be expected to contribute their retained nutrients to the soil where they fall.

In contrast, many other non-seasonal plants, such as palms and conifers, retain their leaves for long periods; *Welwitschia* retains its two main leaves throughout a lifetime that may exceed a thousand years.

The leaf-like organs of Bryophytes (e.g., mosses and liverworts), known as phyllids, differ morphologically from the leaves of vascular plants in that they lack vascular tissue, are usually only a single cell thick and have no cuticle stomata or internal system of intercellular spaces.

Simple, vascularised leaves (microphylls) first evolved as enations, extensions of the stem, in clubmosses such as *Baragwanathia* during the Silurian period. True leaves or euphylls of larger size and with more complex venation did not become widespread in other groups until the Devonian period, by which time the carbon dioxide concentration in the atmosphere had dropped significantly. This occurred independently in several separate lineages of vascular plants, in progymnosperms like *Archaeopteris*, in Sphenopsida, ferns and later in the gymnosperms and angiosperms. Euphylls are also referred to as macrophylls or megaphylls (large leaves).

Morphology (Large-scale Features)

Translucent glands in *Citrus* leaves

A structurally complete leaf of an angiosperm consists of a petiole (leaf stalk), a lamina (leaf blade), and stipules (small structures located to either side of the base of the petiole). Not every species produces leaves with all of these structural components. Stipules may be conspicuous (e.g. beans and roses, soon falling or otherwise not obvious as in Moraceae or absent altogether as in the Magnoliaceae. A petiole may be absent, or the blade may not be laminar (flattened). The tremendous variety shown in leaf structure (anatomy) from species to species is presented in detail below under morphology. The petiole mechanically links the leaf to the plant and provides the route for transfer of water and sugars to and from the leaf. The lamina is typically the location of the majority of photosynthesis. The upper (adaxial) angle between a leaf and a stem is known as the axil of the leaf. It is often the location of a bud. Structures located there are called "axillary".

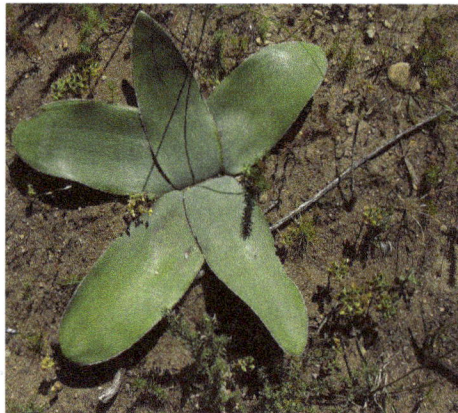

Prostrate leaves in *Crossyne guttata*

External leaf characteristics, such as shape, margin, hairs, the petiole, and the presence of stipules and glands, are frequently important for identifying plants to family, genus or species levels, and botanists have developed a rich terminology for describing leaf characteristics. Leaves almost always have determinate growth. They grow to a specific pattern and shape and then stop. Other plant parts like stems or roots have non-determinate growth, and will usually continue to grow as long as they have the resources to do so.

The type of leaf is usually characteristic of a species (monomorphic), although some species pro-duce more than one type of leaf (dimorphic or polymorphic). The longest leaves are those of the Raffia palm, *R. regalis* which may be up to 25 m (82 ft) long and 3 m (9.8 ft) wide.

Where leaves are basal, and lie on the ground, they are referred to as prostrate.

Basic Leaf Types

Leaves of the White Spruce (*Picea glauca*) are needle-shaped and their arrangement is spiral

- Ferns have fronds

- Conifer leaves are typically needle- or awl-shaped or scale-like

- Angiosperm (flowering plant) leaves: the standard form includes stipules, a petiole, and a lamina

- Lycophytes have microphyll leaves.

- Sheath leaves (type found in most grasses and many other monocots)

- Other specialized leaves (such as those of *Nepenthes*, a pitcher plant)

Arrangement on the Stem

Different terms are usually used to describe the arrangement of leaves on the stem (phyllotaxis):

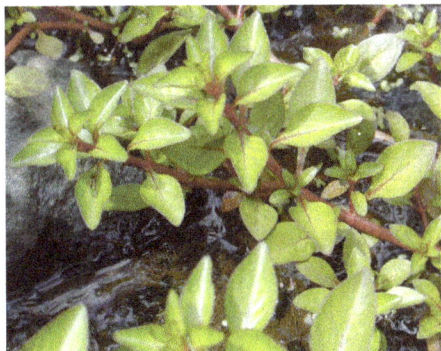

The leaves on this plant are arranged in pairs opposite one another, with successive pairs at right angles to each other (*decussate*) along the red stem. The developing buds in the axils of these leaves.

Alternate

One leaf, branch, or flower part attaches at each point or node on the stem, and leaves alternate direction, to a greater or lesser degree, along the stem.

Basal

Arising from the base of the stem.

Cauline

Arising from the aerial stem.

Opposite

Two leaves, branches, or flower parts attach at each point or node on the stem. Leaf attachments are paired at each node and decussate if, as typical, each successive pair is rotated 90° progressing along the stem.

Whorled, or verticillate

Three or more leaves, branches, or flower parts attach at each point or node on the stem. As with opposite leaves, successive whorls may or may not be decussate, rotated by half the angle between the leaves in the whorl (i.e., successive whorls of three rotated 60°, whorls of four rotated 45°, etc.). Opposite leaves may appear whorled near the tip of the stem. Pseudoverticillate describes an arrangement only appearing whorled, but not actually so.

Rosulate

Leaves form a rosette.

Rows

The term, *distichous*, literally means *two rows*. Leaves in this arrangement may be alternate or opposite in their attachment. The term, *2-ranked*, is equivalent. The terms, *tristichous* and *tetrastichous*, are sometimes encountered. For example, the "leaves" (actually microphylls) of most species of *Selaginella* are tetrastichous, but not decussate.

As a *stem* grows, leaves tend to appear arranged around the stem in a way that optimizes yield of light. In essence, leaves form a helix pattern centered around the stem, either clockwise or counterclockwise, with (depending upon the species) the same angle of divergence. There is a regularity in these angles and they follow the numbers in a Fibonacci sequence: 1/2, 2/3, 3/5, 5/8, 8/13, 13/21, 21/34, 34/55, 55/89. This series tends to the golden angle, which is approximately 360° × 34/89 ≈ 137.52° ≈ 137° 30′. In the series, the numerator indicates the number of complete turns or "gyres" until a leaf arrives at the initial position and the denominator indicates the number of leaves in the arrangement. This can be demonstrated by the following:

- Alternate leaves have an angle of 180° (or 1/2)
- 120° (or 1/3): 3 leaves in 1 circle

- 144° (or 2/5): 5 leaves in 2 gyres
- 135° (or 3/8): 8 leaves in 3 gyres.

Divisions of the Blade

Two basic forms of leaves can be described considering the way the blade (lamina) is divided. A simple leaf has an undivided blade. However, the leaf shape may be formed of lobes, but the gaps between lobes do not reach to the main vein. A compound leaf has a fully subdivided blade, each leaflet of the blade being separated along a main or secondary vein. The leaflets may have petiolules and stipels, the equivalents of the petioles and stipules of leaves. Because each leaflet can appear to be a simple leaf, it is important to recognize where the petiole occurs to identify a compound leaf. Compound leaves are a characteristic of some families of higher plants, such as the Fabaceae. The middle vein of a compound leaf or a frond, when it is present, is called a rachis.

Palmately compound

> Leaves have the leaflets radiating from the end of the petiole, like fingers of the palm of a hand; e.g., *Cannabis* (hemp) and *Aesculus* (buckeyes).

Pinnately compound

> Leaves have the leaflets arranged along the main or mid-vein.

Odd pinnate

> With a terminal leaflet; e.g., *Fraxinus* (ash).

Even pinnate

> Lacking a terminal leaflet; e.g., *Swietenia* (mahogany); A specific type of even pinnate is bipinnate, where leaves only consist of two leaflets, e.g. *Hymenaea*.

Bipinnately compound

> Leaves are twice divided: the leaflets are arranged along a secondary vein that is one of several branching off the rachis. Each leaflet is called a "pinnule". The group of pinnules on each secondary vein forms a "pinna"; e.g., *Albizia* (silk tree).

Trifoliate (or trifoliolate)

> A pinnate leaf with just three leaflets; e.g., *Trifolium* (clover), *Laburnum* (laburnum).

Pinnatifid

> Pinnately dissected to the central vein, but with the leaflets not entirely separate; e.g., *Polypodium*, some *Sorbus* (whitebeams). In pinnately veined leaves the central vein in known as the *midrib*.

Characteristics of the Petiole

Petiolated leaves have a petiole (leaf stalk), and are said to be petiolate.

Sessile (epetiolate) leaves have no petiole and the blade attaches directly to the stem. Subpetiolate leaves are nearly petiolate or have an extremely short petiole and may appear to be sessile.

In clasping or decurrent leaves, the blade partially surrounds the stem.

When the leaf base completely surrounds the stem, the leaves are said to be perfoliate, such as in *Eupatorium perfoliatum*.

The overgrown petioles of rhubarb (*Rheum rhabarbarum*) are edible.

In peltate leaves, the petiole attaches to the blade inside the blade margin.

In some *Acacia* species, such as the koa tree (*Acacia koa*), the petioles are expanded or broadened and function like leaf blades; these are called phyllodes. There may or may not be normal pinnate leaves at the tip of the phyllode.

A stipule, present on the leaves of many dicotyledons, is an appendage on each side at the base of the petiole, resembling a small leaf. Stipules may be lasting and not be shed (a stipulate leaf, such as in roses and beans), or be shed as the leaf expands, leaving a stipule scar on the twig (an exstipulate leaf). The situation, arrangement, and structure of the stipules is called the "stipulation".

Free, lateral

> As in *Hibiscus*.

Adnate

> Fused to the petiole base, as in *Rosa*.

Ochreate

> Provided with ochrea, or sheath-formed stipules, as in Polygonaceae; e.g., rhubarb.

Encircling the petiole base

> *Interpetiolar*

> Between the petioles of two opposite leaves, as in Rubiaceae.

> *Intrapetiolar*

> Between the petiole and the subtending stem, as in Malpighiaceae.

Veins

Veins (sometimes referred to as nerves) constitute one of the more visible leaf traits or characteristics. The veins in a leaf represent the vascular structure of the organ, extending into the leaf via the petiole and provide transportation of water and nutrients between leaf and stem, and play a crucial role in the maintenance of leaf water status and photosynthetic capacity. They also play a role in the mechanical support of the leaf. Within the lamina of the leaf, while some vascular plants possess only a single vein, in most this vasculature generally divides (ramifies) according to a variety of patterns (venation) and form cylindrical bundles, usually lying in the median plane of the mesophyll, between the two layers of epidermis. This pattern is often specific to taxa, and of which angiosperms possess two main types, parallel and reticulate (net like). In general, parallel venation is typical of monocots, while reticulate is more typical of eudicots and magnoliids ("dicots"), though there are many exceptions.

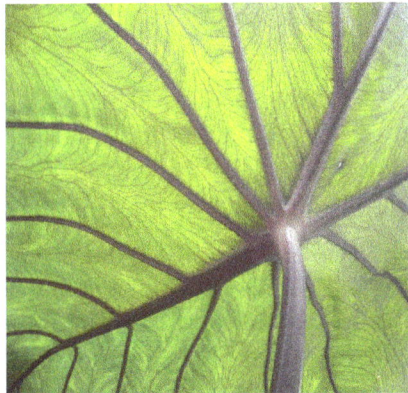
Branching veins on underside of taro leaf

The vein or veins entering the leaf from the petiole are called primary or first order veins. The veins branching from these are secondary or second order veins. These primary and secondary veins are considered major veins or lower order veins, though some authors include third order. Each subsequent branching is sequentially numbered, and these are the higher order veins, each branching being associated with a narrower vein diameter. In parallel veined leaves, the primary veins run parallel and equidistant to each other for most of the length of the leaf and then converge or fuse (anastomose) towards the apex. Usually many smaller minor veins interconnect these primary veins, but may terminate with very fine vein endings in the mesophyll. Minor veins are more typical of angiosperms, which may have as many as four higher orders. In contrast, leaves with reticulate venation there is a single (sometimes more) primary vein in the centre of the leaf, referred to as the midrib or costa and is continuous with the vasculature of the petiole more proximally. The midrib then branches to a number of smaller secondary veins, also known as second order veins, that extend toward the leaf margins. These often terminate in a hydathode, a secretory organ, at the margin. In turn, smaller veins branch from the secondary veins, known as tertiary or third order (or higher order) veins, forming a dense reticulate pattern. The areas or islands of mesophyll lying between the higher order veins, are called areoles. Some of the smallest veins (veinlets) may have their endings in the areoles, a process known as areolation. These minor veins act as the sites of exchange between the mesophyll and the plant's vascular system. Thus minor veins collect the products of photosynthesis (photosynthate) from the cells where it takes place, while major veins are responsible for its transport outside of the leaf. At the same time water is being transported in the opposite direction.

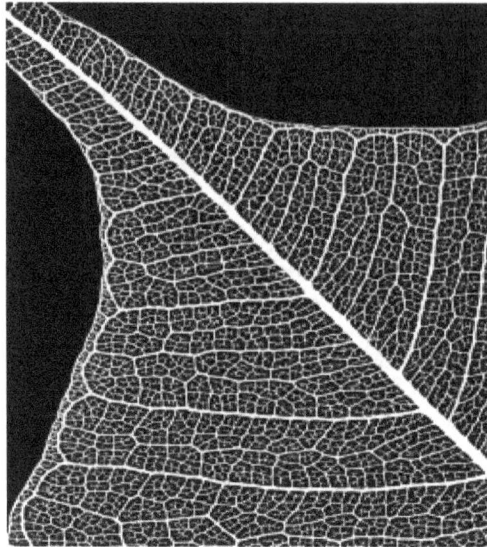

Micrograph of a leaf skeleton

The number of vein endings is very variable, as is whether second order veins end at the margin, or link back to other veins. There are many elaborate variations on the patterns that the leaf veins form, and these have functional implications. Of these, angiosperms have the greatest diversity. Within these the major veins function as the support and distribution network for leaves and are correlated with leaf shape. For instance the parallel venation found in most monocots correlates with their elongated leaf shape and wide leaf base, while reticulate venation is seen in simple entire leaves, while digitate leaves typically have venation in which three or more primary veins diverge radially from a single point.

In evolutionary terms, early emerging taxa tend to have dichotomous branching with reticulate systems emerging later. Veins appeared in the Permian period (299–252 mya), prior to the appearance of angiosperms in the Triassic (252–201 mya), during which vein hierarchy appeared enabling higher function, larger leaf size and adaption to a wider vaiety of climatic conditions. Although it is the more complex pattern, branching veins appear to be plesiomorphic and in some form were present in ancient seed plants as long as 250 million years ago. A pseudo-reticulate venation that is actually a highly modified penniparallel one is an autapomorphy of some Melanthiaceae, which are monocots; e.g., *Paris quadrifolia* (True-lover's Knot). In leaves with reticulate venation, veins form a scaffolding matrix imparting mechanical rigidity to leaves.

Morphology Changes within a Single Plant

Homoblasty

> Characteristic in which a plant has small changes in leaf size, shape, and growth habit between juvenile and adult stages, in contrast to;

Heteroblasty

> Characteristic in which a plant has marked changes in leaf size, shape, and growth habit between juvenile and adult stages.

Anatomy (Medium and Small Scale)

Medium-scale Features

Leaves are normally extensively vascularised and typically have networks of vascular bundles containing xylem, which supplies water for photosynthesis, and phloem, which transports the sugars produced by photosynthesis. Many leaves are covered in trichomes (small hairs) which have diverse structures and functions.

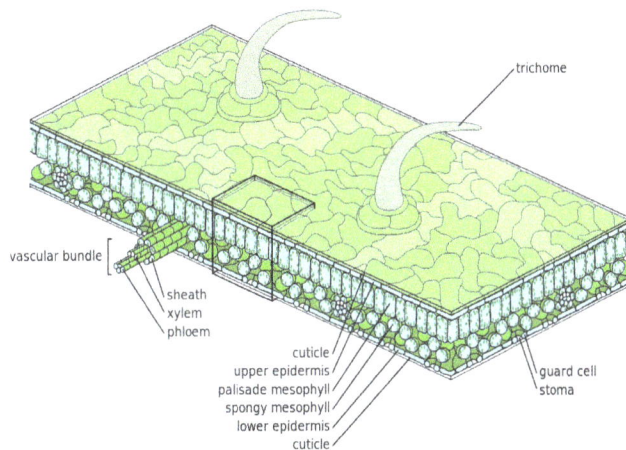

Small-scale Features

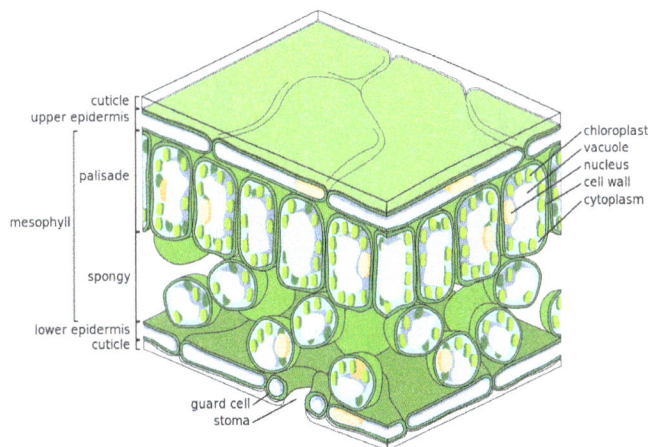

The major tissue systems present are

1. The epidermis, which covers the upper and lower surfaces

2. The mesophyll tissue inside the leaf, which is rich in chloroplasts (also called chlorenchyma)

3. The arrangement of veins (the vascular tissue)

These three tissue systems typically form a regular organisation at the cellular scale. Specialised cells that differ markedly from surrounding cells, and which often synthesise specialised products such as crystals, are termed idioblasts.

Major Leaf Tissues

Cross-section of a leaf

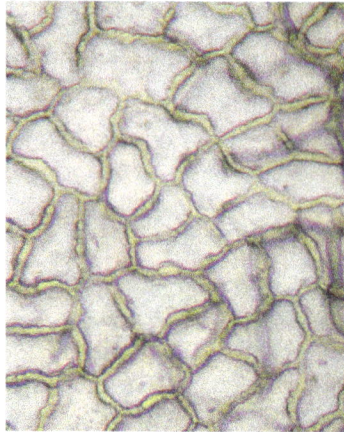
Epidermal cells

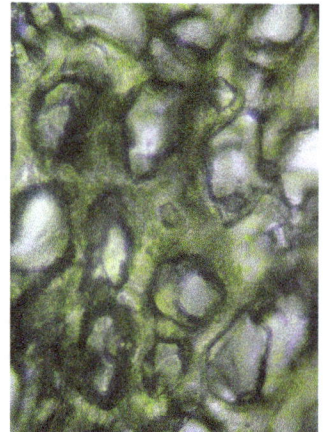
Spongy mesophyll cells

Epidermis

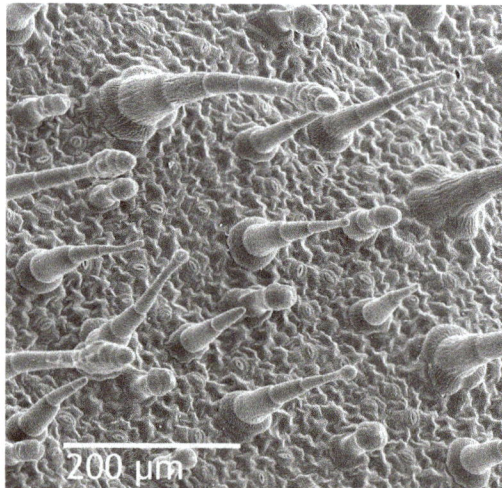

SEM image of the leaf epidermis of *Nicotiana alata*, showing trichomes
(hair-like appendages) and stomata (eye-shaped slits, visible at full resolution).

The epidermis is the outer layer of cells covering the leaf. It is covered with a waxy cuticle which is impermeable to liquid water and water vapor and forms the boundary separating the plant's inner cells from the external world. The cuticle is in some cases thinner on the lower epidermis than on the upper epidermis, and is generally thicker on leaves from dry climates as compared with those from wet climates. The epidermis serves several functions: protection against water loss by way of transpiration, regulation of gas exchange and secretion of metabolic compounds. Most leaves show dorsoventral anatomy: The upper (adaxial) and lower (abaxial) surfaces have somewhat different construction and may serve different functions.

The epidermis tissue includes several differentiated cell types; epidermal cells, epidermal hair cells (trichomes), cells in the stomatal complex; guard cells and subsidiary cells. The epidermal cells are the most numerous, largest, and least specialized and form the majority of the epidermis. They are typically more elongated in the leaves of monocots than in those of dicots.

Chloroplasts are generally absent in epidermal cells, the exception being the guard cells of the stomata. The stomatal pores perforate the epidermis and are surrounded on each side by chloroplast-containing guard cells, and two to four subsidiary cells that lack chloroplasts, forming a specialized cell group known as the stomatal complex. The opening and closing of the stomatal aperture is controlled by the stomatal complex and regulates the exchange of gases and water vapor between the outside air and the interior of the leaf. Stomata therefore play the important role in allowing photosynthesis without letting the leaf dry out. In a typical leaf, the stomata are more numerous over the abaxial (lower) epidermis than the adaxial (upper) epidermis and are more numerous in plants from cooler climates.

Mesophyll

Most of the interior of the leaf between the upper and lower layers of epidermis is a *parenchyma* (ground tissue) or *chlorenchyma* tissue called the mesophyll (Greek for "middle leaf"). This assimilation tissue is the primary location of photosynthesis in the plant. The products of photosynthesis are called "assimilates".

In ferns and most flowering plants, the mesophyll is divided into two layers:

- An upper palisade layer of vertically elongated cells, one to two cells thick, directly beneath the adaxial epidermis, with intercellular air spaces between them. Its cells contain many more chloroplasts than the spongy layer. These long cylindrical cells are regularly arranged in one to five rows. Cylindrical cells, with the *chloroplasts* close to the walls of the cell, can take optimal advantage of light. The slight separation of the cells provides maximum absorption of carbon dioxide. Sun leaves have a multi-layered palisade layer, while shade leaves or older leaves closer to the soil are single-layered.

- Beneath the palisade layer is the spongy layer. The cells of the spongy layer are more branched and not so tightly packed, so that there are large intercellular air spaces between them for oxygen and carbon dioxide to diffuse in and out of during respiration and photosynthesis. These cells contain fewer chloroplasts than those of the palisade layer. The pores or *stomata* of the epidermis open into substomatal chambers, which are connected to the intercellular air spaces between the spongy and palisade mesophyll cells.

Leaves are normally green, due to chlorophyll in chloroplasts in the mesophyll cells. Plants that lack chlorophyll cannot photosynthesize.

Vascular Tissue

The veins are the vascular tissue of the leaf and are located in the spongy layer of the mesophyll. The pattern of the veins is called venation. In angiosperms the venation is typically parallel in monocotyledons and forms an interconnecting network in broad-leaved plants. They were once thought to be typical examples of pattern formation through ramification, but they may instead exemplify a pattern formed in a stress tensor field.

A vein is made up of a vascular bundle. At the core of each bundle are clusters of two distinct types of conducting cells:

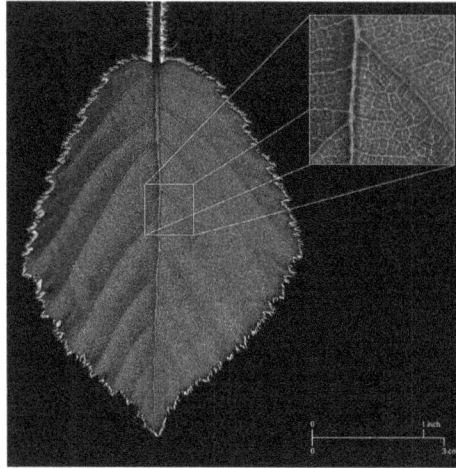

The veins of a bramble leaf

- Xylem: cells that bring water and minerals from the roots into the leaf.

- Phloem: cells that usually move sap, with dissolved sucrose(glucose to sucrose) produced by photosynthesis in the leaf, out of the leaf.

The xylem typically lies on the adaxial side of the vascular bundle and the phloem typically lies on the abaxial side. Both are embedded in a dense parenchyma tissue, called the sheath, which usually includes some structural collenchyma tissue.

Leaf Development

According to Agnes Arber's partial-shoot theory of the leaf, leaves are partial shoots, being derived from leaf primordia of the shoot apex. Compound leaves are closer to shoots than simple leaves. Developmental studies have shown that compound leaves, like shoots, may branch in three dimensions. On the basis of molecular genetics, Eckardt and Baum (2010) concluded that "it is now generally accepted that compound leaves express both leaf and shoot properties."

Ecology

Biomechanics

Plants respond and adapt to environmental factors, such as light and mechanical stress from wind. Leaves need to support their own mass and align themselves in such a way as to optimise their exposure to the sun, generally more or less horizontally. However horizontal alignment maximises exposure to bending forces and failure from stresses such as wind, snow, hail, falling debris, animals, and abrasion from surrounding foliage and plant structures. Overall leaves are relatively flimsy with regard to other plant structures such as stems, branches and roots.

Both leaf blade and petiole structure influence the leaf's response to forces such as wind, allowing a degree of repositioning to minimise drag and damage, as opposed to resistance. Such leaf movement may also increase turbulence of the air close to the surface of the leaf, which thins the boundary layer of air immediately adjacent to the surface, increasing the capacity for gas and heat exchange, as well as photosynthesis. Strong wind forces may result in diminished leaf number and surface area, which while reducing drag, involves a trade off of also reduces photosynthesis. Thus, leaf design may involve compromise between carbon gain, thermoregulation and water loss on the one hand, and the cost of sustaining both static and dynamic loads. In vascular plants, perpendicular forces are spread over a larger area and are relatively flexible in both bending and torsion, enabling elastic deforming without damage.

Many leaves rely on hydrostatic support arranged around a skeleton of vascular tissue for their strength, which depends on maintaining leaf water status. Both the mechanics and architecture of the leaf reflect the need for transportation and support. Read and Stokes (2006) consider two basic models, the "hydrostatic" and "I-beam leaf" form. Hydrostatic leaves such as in *Prostanthera lasianthos* are large and thin, and may involve the need for multiple leaves rather single large leaves because of the amount of veins needed to support the periphery of large leaves. But large leaf size favours efficiency in photosynthesis and water conservation, involving further trade offs. On the other hand I-beam leaves such as *Banksia marginata* involve specialised structures to stiffen them. These I-beams are formed from bundle sheath extensions of sclerenchyma meeting stiffened sub-epidermal layers. This shifts the balance from reliance on hydrostatic pressure to structural support, an obvious advantage where water is relatively scarce. Long narrow leaves bend more easily than ovate leaf blades of the same area. Monocots typically have such linear leaves that maximise surface area while minimising self-shading. In these a high proportion of longitudinal main veins provide additional support.

Interactions with other Organisms

Some insects, like *Kallima inachus*, mimic leaves

Although not as nutritious as other organs such as fruit, leaves provide a food source for many organisms. The leaf is a vital source of energy production for the plant, and plants have evolved protection against animals that consume leaves, such as tannins, chemicals which hinder the digestion of proteins and have an unpleasant taste. Animals that are specialized to eat leaves are known as folivores.

Some species have cryptic adaptations by which they use leaves in avoiding predators. For example, the caterpillars of some leaf-roller moths will create a small home in the leaf by folding it over themselves. Some sawflies similarly roll the leaves of their food plants into tubes. Females of the Attelabidae, so-called leaf-rolling weevils, lay their eggs into leaves that they then roll up as means of protection. Other herbivores and their predators mimic the appearance of the leaf. Reptiles such as some chameleons, and insects such as some katydids, also mimic the oscillating movements of leaves in the wind, moving from side to side or back and forth while evading a possible threat.

Seasonal Leaf Loss

Leaves shifting color in autumn (fall)

Leaves in temperate, boreal, and seasonally dry zones may be seasonally deciduous (falling off or dying for the inclement season). This mechanism to shed leaves is called abscission. When the leaf is shed, it leaves a leaf scar on the twig. In cold autumns, they sometimes change color, and turn yellow, bright-orange, or red, as various accessory pigments (carotenoids and xanthophylls) are revealed when the tree responds to cold and reduced sunlight by curtailing chlorophyll production. Red anthocyanin pigments are now thought to be produced in the leaf as it dies, possibly to mask the yellow hue left when the chlorophyll is lost—yellow leaves appear to attract herbivores such as aphids. Optical masking of chlorophyll by anthocyanins reduces risk of photo-oxidative damage to leaf cells as they senesce, which otherwise may lower the efficiency of nutrient retrieval from senescing autumn leaves.

Evolutionary Adaptation

In the course of evolution, leaves have adapted to different environments in the following ways:

- Waxy micro- and nanostructures on the surface reduce wetting by rain and adhesion of contamination.

- Divided and compound leaves reduce wind resistance and promote cooling.

- Hairs on the leaf surface trap humidity in dry climates and create a boundary layer reducing water loss.

- Waxy plant cuticles reduce water loss.

- Large surface area provides a large area for capture of sunlight.

- In harmful levels of sunlight, specialised leaves, opaque or partly buried, admit light through a translucent leaf window for photosynthesis at inner leaf surfaces (e.g. *Fenestraria*).

- Succulent leaves store water and organic acids for use in CAM photosynthesis.

- Aromatic oils, poisons or pheromones produced by leaf borne glands deter herbivores (e.g. eucalypts).

- Inclusions of crystalline minerals deter herbivores (e.g. silica phytoliths in grasses, raphides in Araceae).

- Petals attract pollinators.

- Spines protect the plants from herbivores (e.g. cacti).

- Stinging hairs to protect against herbivory, e.g. in *Urtica dioica* and *Dendrocnide moroides* (Urticaceae).

- Special leaves on carnivorous plants are adapted for trapping food, mainly invertebrate prey, though some species trap small vertebrates as well.

- Bulbs store food and water (e.g. onions).

- Tendrils allow the plant to climb (e.g. peas).

- Bracts and pseudanthia (false flowers) replace normal flower structures when the true flowers are greatly reduced (e.g. spurges and spathes in the Araceae.

Shape

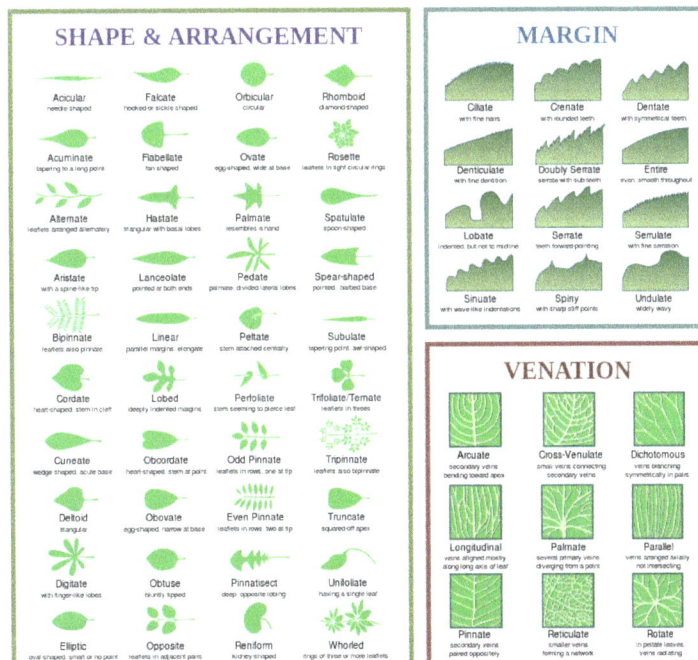

Leaf morphology terms

Leaves showing various morphologies. Clockwise from upper left: tripartite lobation, elliptic with serrulate margin, palmate venation, acuminate odd-pinnate (center), pinnatisect, lobed, elliptic with entire margin

Edge (Margin)

Term	Description
Entire	Even; with a smooth margin; without toothing
Ciliate	Fringed with hairs
Crenate	Wavy-toothed; dentate with rounded teeth
Dentate	Toothed May be coarsely dentate, having large teeth or glandular dentate, having teeth which bear glands
Denticulate	Finely toothed
Doubly serrate	Each tooth bearing smaller teeth
Serrate	Saw-toothed; with asymmetrical teeth pointing forward
Serrulate	Finely serrate
Sinuate	With deep, wave-like indentations; coarsely crenate
Lobate	Indented, with the indentations not reaching the center
Undulate	With a wavy edge, shallower than sinuate
Spiny or pungent	With stiff, sharp points such as thistles

Apex (Tip)

Term	Description
Acuminate	Long-pointed, prolonged into a narrow, tapering point in a concave manner
Acute	Ending in a sharp, but not prolonged point
Cuspidate	With a sharp, elongated, rigid tip; tipped with a cusp
Emarginate	Indented, with a shallow notch at the tip
Mucronate	Abruptly tipped with a small short point
Mucronulate	Mucronate, but with a noticeably diminutive spine
Obcordate	Inversely heart-shaped
Obtuse	Rounded or blunt
Truncate	Ending abruptly with a flat end

Classification

Hickey Primary Venation Types

Pinnate venation, Ostrya virginiana

Parallel venation, *Iris*

Campylodromous venation, *Maianthemum bifolium*

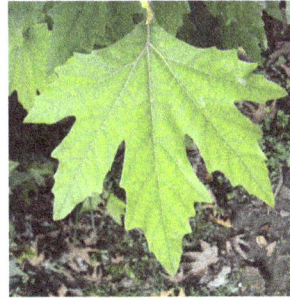

Palinactodromous venation, *Platanus orientalis*

A number of different classification systems of the patterns of leaf veins (venation or veination) have been described, starting with Ettingshausen (1861), together with many different descriptive terms, and the terminology has been described as "formidable". One of the commonest among these is the Hickey system, originally developed for "dicotyledons" and using a number of Ettingshausen's terms derived from Greek.

Hickey System

1. Pinnate (feather-veined, reticulate, pinnate-netted, penniribbed, penninerved, or penniveined)

The veins arise pinnately (feather like) from a single primary vein (mid-vein) and subdivide into secondary veinlets, known as higher order veins. These, in turn, form a complicated network. This type of venation is typical for (but by no means limited to) "dicotyledons" (non monocotyledon angiosperms). E.g. *Ostrya* There are three subtypes of pinnate venation;

- *Craspedodromous* (Greek: *kraspedon* - edge, *dromos* - running): The major veins reach to the margin of the leaf.

- *Camptodromous*: Major veins extend close to the margin, but bend before they intersect with the margin.

- *Hyphodromous*: all secondary veins are absent, rudimentary or concealed

These in turn have a number of further subtypes such as eucamptodromous, where secondary veins curve near the margin without joining adjacent secondary veins.

Pinnate

Craspedodromous   Camptodromous   Hyphodromous

2. Parallelodromous (Parallel-veined, parallel-ribbed, parallel-nerved, penniparallel, striate)

Two or more primary veins originating beside each other at the leaf base, and running parallel to each other to the apex and then converging there. Commissural veins (small veins) connect the major parallel veins. Typical for most monocotyledons, such as grasses. The additional terms marginal (primary veins reach the margin), and reticulate (primary veins do not reach the margin) are also used.

Parallelodromous

Campylodromous (*campylos* - curve)

Several primary veins or branches originating at or close to a single point and running in recurved arches, then converging at apex. E.g. *Maianthemum*

Campylodromous

Acrodromous

Two or more primary or well developed secondary veins in convergent arches towards apex, without basal recurvature as in Campylodromous. May be basal or suprabasal depending on origin, and perfect or imperfect depending on whether they reach to 2/3 of the way to the apex. E.g. *Miconia* (basal type), *Endlicheria* (suprabasal type)

Acrodromous

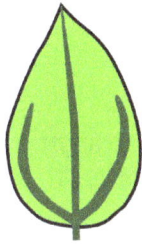

Imperfect basal   Imperfect suprabasal   Perfect basal   Perfect suprabasal

5. Actinodromous

Three or more primary veins diverging radially from a single point. E.g. *Arcangelisia* (basal type), *Givotia* (suprabasal type)

Actinodromous

Imperfect marginal   Imperfect reticulate

6. Palinactodromous

Primary veins with one or more points of secondary dichotomous branching beyond the primary divergence, either closely or more distantly spaced. E.g. *Platanus*

Palinactodromous

Types 4–6 may similarly be subclassified as basal (primaries joined at the base of the blade) or suprabasal (diverging above the blade base), and perfect or imperfect, but also flabellate.

At about the same time, Melville (1976) described a system applicable to all Angiosperms and using Latin and English terminology. Melville also had six divisions, based on the order in which veins develop.

- Arbuscular (arbuscularis) - branching repeatedly by regular dichotomy to give rise to a three dimensional bush-like structure consisting of linear segment (2 subclasses)

- Flabellate (flabellatus) - primary veins straight or only slightly curved, diverging from the base in a fan-like manner (4 subclasses)

- Palmate (palmatus) - curved primary veins (3 subclasses)

- Pinnate (pinnatus) - single primary vein, the midrib, along which straight or arching secondary veins are arranged at more or less regular intervals (6 subclasses)

- Collimate (collimatus) - numerous longitudinally parallel primary veins arising from a transverse meristem (5 subclasses)

- Conglutinate (conglutinatus) - derived from fused pinnate leaflets (3 subclasses)

A modified form of the Hickey system was later incorporated into the Smithsonian classification (1999) which proposed seven main types of venation, based on the architecture of the primary veins, adding Flabellate as an additional main type. Further classification was then made on the basis of secondary veins, with 12 further types, such as;

- *brochidodromous* (closed form in which the secondaries are joined together in a series of prominent arches, as in *Hildegardia*)

- *craspedodromous* (open form with secondaries terminating at the margin, in toothed leaves, as in *Celtis*)

- *eucamptodromous* (intermediate form with upturned secondaries that gradually diminish apically but inside the margin, and connected by intermediate tertiary veins rather than loops between secondaries, as in *Cornus*)

- *cladodromous* (secondaries freely branching toward the margin, as in *Rhus*)

terms which had been used as subtypes in the original Hickey system.

Secondary Venation Patterns

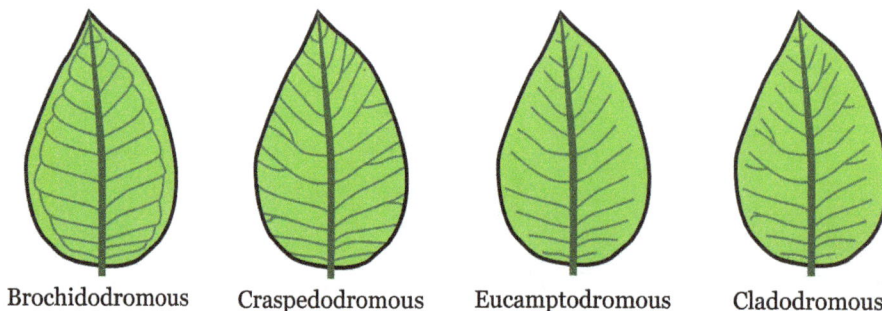

Brochidodromous   Craspedodromous   Eucamptodromous   Cladodromous

Flabellate

Analyses of vein patterns often fall into consideration of the vein orders, primary vein type, secondary vein type (major veins), and minor vein density. A number of authors have adopted simplified versions of these schemes. At its simplest the primary vein types can be considered in three or four groups depending on the plant divisions being considered;

- pinnate

- palmate

- parallel

where palmate refers to multiple primary veins that radiate from the petiole, as opposed to branching from the central main vein in the pinnate form, and encompasses both of Hickey types 4 and 5, which are preserved as subtypes, e.g. palmate-acrodromous.

Palmate, Palmate-netted, Palmate-veined, Fan-veined

Several main veins of approximately equal size diverge from a common point near the leaf base where the petiole attaches, and radiate toward the edge of the leaf. Palmately veined leaves are often lobed or divided with lobes radiating from the common point. They may vary in the number of primary veins (3 or more), but always radiate from a common point. e.g. most *Acer* (maples).

Other Systems

Alternatively, Simpson uses:

- uninervous - central midrib with no lateral veins (microphyllous), seen in the non-seed bearing tracheophytes, such as horsetails

- dichotomous - veins successively branching into equally sized veins from a common point, forming a Y junction, fanning out. Amongst temperate woody plants, *Ginkgo biloba* is the only species exhibiting dichotomous venation. Also some pteridophytes (ferns).

- parallel - primary and secondary veins roughly parallel to each other, running the length of the leaf, often connected by short perpendicular links, rather than form networks. In some species, the parallel veins join together at the base and apex, such as needle-type evergreens and grasses. Characteristic of monocotyledons, but exceptions include *Arisaema*, and as below, under netted.

- netted (reticulate, pinnate) - a prominent midvein with secondary veins branching off along both sides of it. The name derives from the ultimate veinlets which form an interconnecting net like pattern or network. (The primary and secondary venation may be referred to as pinnate, while the net like finer veins are referred to as netted or reticulate); most non-monocot angiosperms, exceptions including *Calophyllum*. Some monocots have reticulate venation, including *Colocasia*, *Dioscorea* and *Smilax*.

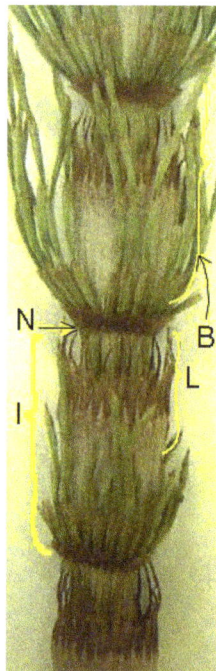

Equisetum: Reduced microphyllous leaves (L) arising in whorl from node

However, these simplified systems allow for further division into multiple subtypes. Simpson, (and others) divides parallel and netted (and some use only these two terms for Angiosperms) on the basis of the number of primary veins (costa) as follows;

Ginkgo biloba:
Dichotomous venation

- Parallel

    o Penni-parallel (pinnate, pinnate parallel, unicostate parallel) - single central prominent midrib, secondary veins from this arise perpendicularly to it and run parallel to each other towards the margin or tip, but do not join (anastomose). The term unicostate refers to the prominence of the single midrib (costa) running the length of the leaf from base to apex. e.g. Zingiberales, such as Bananas etc.

    o Palmate-parallel (multicostate parallel) - several equally prominent primary veins arising from a single point at the base and running parallel towards tip or margin. The term multicostate refers to having more than one prominent main vein. e.g. "fan" (palmate) palms (Arecaceae)

        • Multicostate parallel convergent - mid-veins converge at apex e.g. *Bambusa arundinacea = B. bambos* (Aracaceae), *Eichornia*

        • Multicostate parallel divergent - mid-veins diverge more or less parallel towards the margin e.g. *Borassus* (Poaceae), fan palms

- Netted (Reticulate)

    o Pinnately (veined, netted, unicostate reticulate) - Single prominent midrib running from base to apex, secondary veins arising on both sides along the length of the primary midrib, running towards the margin or apex (tip), with a network of smaller veinlets forming a reticulum (mesh or network). e.g. *Mangifera, Ficus religiosa, Psidium guajava, Hibiscus rosa-sinensis, Salix alba*

    o Palmately (multicostate reticulate) - more than one primary veins arising from a single point, running from base to apex. e.g. *Liquidambar styraciflua* This may be further subdivided;

        • Multicostate convergent - major veins diverge from origin at base then converge towards the tip. e.g. *Zizyphus, Smilax, Cinnamomum*

- Multicostate divergent - all major veins diverge towards the tip. e.g. *Gossypium, Cucurbita, Carica papaya, Ricinus communis*

  o Ternately (ternate-netted) - three primary veins, as above, e.g. *Ceanothus leucodermis, C. tomentosus, Encelia farinosa*

Simpson Venation Patterns

Maranta leuconeura var. *erythro-neura* (Zingiberales): Penni-parallel

Coccothrinax argentea (Arecaceae): Palmate-parallel

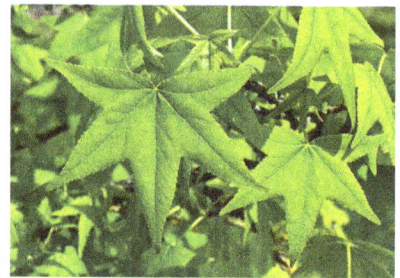

Liquidambar styraciflua: Palmately netted

These complex systems are not used much in morphological descriptions of taxa, but have usefulness in plant identification, although criticized as being unduly burdened with jargon.

Ziziphus jujuba: Multicostate palmate convergent

An older, even simpler system, used in some flora uses only two categories, open and closed.

- Open: Higher order veins have free endings among the cells and are more characteristic of non-monocotyledon angiosperms. They are more likely to be associated with leaf shapes that are toothed, lobed or compound. They may be subdivided as;

  o Pinnate (feather-veined) leaves, with a main central vein or rib (midrib), from which the remainder of the vein system arises

  o Palmate, in which three or more main ribs rise together at the base of the leaf, and diverge upward.

  o Dichotomous, as in ferns, where the veins fork repeatedly

- Closed: Higher order veins are connected in loops without ending freely among the cells. These tend to be in leaves with smooth outlines, and are characteristic of monocotyledons.

o   They may be subdivided into whether the veins run parallel, as in grasses, or have other patterns.

Other Descriptive Terms

There are also many other descriptive terms, often with very specialised usage and confined to specific taxonomic groups. The conspicuousness of veins depends on a number of features. These include the width of the veins, their prominence in relation to the lamina surface and the degree of opacity of the surface, which may hide finer veins. In this regard, veins are called obscure and the order of veins that are obscured and whether upper, lower or both surfaces, further specified.

Terms that describe vein prominence include bullate, channelled, flat, guttered, impressed, prominent and recessed (Hawthorne & Lawrence 2013). Veins may show different types of prominence in different areas of the leaf. For instance *Pimenta racemosa* has a channelled midrib on the upper surfae, but this is prominent on the lower surface.

Vein prominence:

Bullate

Surface of leaf raised in a series of domes between the veins on the upper surface, and therefore also with marked depressions. e.g. *Rytigynia pauciflora*, *Vitis vinifera*

Channelled (canalicululate)

Veins sunken below the surface, resulting in a rounded channel. Sometimes confused with "guttered" because the channels may function as gutters for rain to run off and allow drying, as in many Melastomataceae. e.g. *Pimenta racemosa* (Myrtaceae), *Clidemia hirta* (Melastomataceae).

Guttered

Veins partly prominent, the crest above the leaf lamina surface, but with channels running along each side, like gutters

Impressed

Vein forming raised line or ridge which lies below the plane of the surface which bears it, as if pressed into it, and are often exposed on the lower surface. Tissue near the veins often appears to pucker, giving them a sunken or embossed appearance

Obscure

Veins not visible, or not at all clear; if unspecified, then not visible with the naked eye. e.g. *Berberis gagnepainii*. In this *Berberis*, the veins are only obscure on the undersurface.

Prominent

Vein raised above surrounding surface so to be easily felt when stroked with finger. e.g.

Recessed

Vein is sunk below the surface, more prominent than surrounding tissues but more sunken in channel than with impressed veins. e.g. *Viburnum plicatum*.

Types of Vein Prominence

| *Vitis vinifera* Bullate | *Clidemia hirta* Channeled | *Berberis gagnepainii* Obscure (under surface) |

Other:

Plinervy (plinerved)

More than one main vein (nerve) at the base. Lateral secondary veins branching from a point above the base of the leaf. Usually expressed as a suffix, as in 3-plinerved or triplinerved leaf. In a 3-plinerved (triplinerved) leaf three main veins branch above the base of the lamina (two secondary veins and the main vein) and run essentially parallel subsequently, as in *Ceanothus* and in *Celtis*. Similarly a quintuplinerve (five-veined) leaf has four secondary veins and a main vein. A pattern with 3-7 veins is especially conspicuous in Melastomataceae. The term has also been used in Vaccinieae. The term has been used as synonymous with acrodromous, palmate-acrodromous or suprabasal acrodromous, and is thought to be too broadly defined.

Scalariform

Veins arranged like the rungs of a ladder, particularly higher order veins

Submarginal

veins running close to leaf margin

Trinerved

2 major basal nerves besides the midrib

Diagrams of venation patterns

Term	Description
Arcuate	Secondary arching toward the apex
Dichotomous	Veins splitting in two
Longitudinal	All veins aligned mostly with the midvein

Term	Description
Parallel	All veins parallel and not intersecting
Pinnate	Secondary veins borne from midrib
Reticulate	All veins branching repeatedly, net veined
Rotate	Veins coming from the center of the leaf and radiating toward the edges
Transverse	Tertiary veins running perpendicular to axis of main vein, connecting secondary veins

Size

The terms megaphyll, macrophyll, mesophyll, notophyll, microphyll, nanophyll and leptophyll are used to describe leaf sizes (in descending order), in a classification devised in 1934 by Christen C. Raunkiær and since modified by others.

Seed

Brown flax seeds

A seed is an embryonic plant enclosed in a protective outer covering. The formation of the seed is part of the process of reproduction in seed plants, the spermatophytes, including the gymnosperm and angiosperm plants.

Seeds are the product of the ripened ovule, after fertilization by pollen and some growth within the mother plant. The embryo is developed from the zygote and the seed coat from the integuments of the ovule.

Seeds have been an important development in the reproduction and success of gymnosperm and angiosperm plants, relative to more primitive plants such as ferns, mosses and liverworts, which do not have seeds and use water-dependent means to propagate themselves. Seed plants now dominate biological niches on land, from forests to grasslands both in hot and cold climates.

The term "seed" also has a general meaning that antedates the above—anything that can be sown, e.g. "seed" potatoes, "seeds" of corn or sunflower "seeds". In the case of sunflower and corn "seeds", what is sown is the seed enclosed in a shell or husk, whereas the potato is a tuber.

Many structures commonly referred to as "seeds" are actually dry fruits. Plants producing berries are called baccate. Sunflower seeds are sometimes sold commercially while still enclosed within the hard wall of the fruit, which must be split open to reach the seed. Different groups of plants have other modifications, the so-called stone fruits (such as the peach) have a hardened fruit layer (the endocarp) fused to and surrounding the actual seed. Nuts are the one-seeded, hard-shelled fruit of some plants with an indehiscent seed, such as an acorn or hazelnut.

Production

Seeds are produced in several related groups of plants, and their manner of production distinguishes the angiosperms ("enclosed seeds") from the gymnosperms ("naked seeds"). Angiosperm seeds are produced in a hard or fleshy structure called a fruit that encloses the seeds, hence the name. Some fruits have layers of both hard and fleshy material. In gymnosperms, no special structure develops to enclose the seeds, which begin their development "naked" on the bracts of cones. However, the seeds do become covered by the cone scales as they develop in some species of conifer.

Seed production in natural plant populations varies widely from year to year in response to weather variables, insects and diseases, and internal cycles within the plants themselves. Over a 20-year period, for example, forests composed of loblolly pine and shortleaf pine produced from 0 to nearly 5 million sound pine seeds per hectare. Over this period, there were six bumper, five poor, and nine good seed crops, when evaluated for production of adequate seedlings for natural forest reproduction.

Development

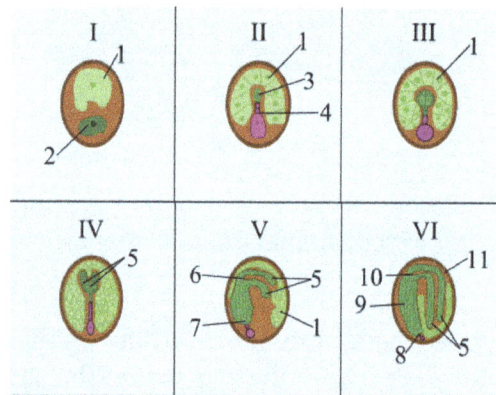

Stages of seed development:
I Zygote II Proembryo III Globular IV Heart V Torpedo VI Mature Embryo
Key: *1. Endosperm 2. Zygote 3. Embryo 4. Suspensor 5. Cotyledons 6. Shoot Apical Meristem 7. Root Apical Meristem 8. Radicle 9. Hypocotyl 10. Epicotyl 11. Seed Coat*

Angiosperm (flowering plants) seeds consist of three genetically distinct constituents: (1) the embryo formed from the zygote, (2) the endosperm, which is normally triploid, (3) the seed coat from tissue derived from the maternal tissue of the ovule. In angiosperms, the process of seed development begins with double fertilization, which involves the fusion of two male gametes with the egg cell and the central cell to form the primary endosperm and the zygote. Right after fertilization, the zygote is mostly inactive, but the primary endosperm divides rapidly to form the endosperm tissue. This tissue becomes the food the young plant will consume until the roots have developed after germination.

Ovule

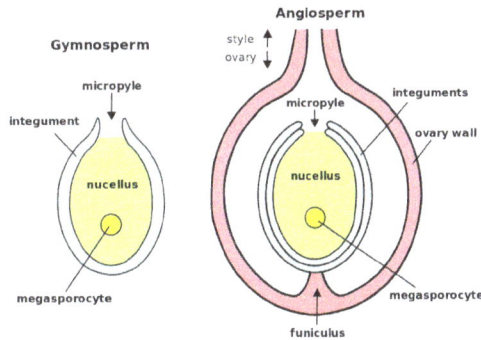

Plant ovules: Gymnosperm ovule on left, angiosperm ovule (inside ovary) on right

After fertilization the ovules develop into the seeds. The ovule consists of a number of components:

- The funicle (*funiculus, funiculi*) or seed stalk which attaches the ovule to the placenta and hence ovary or fruit wall, at the pericarp.

- The nucellus, the remnant of the megasporangium and main region of the ovule where the megagametophyte develops.

- The micropyle, a small pore or opening in the apex of the integument of the ovule where the pollen tube usually enters during the process of fertilization.

- The chalaza, the base of the ovule opposite the micropyle, where integument and nucellus are joined together).

The shape of the ovules as they develop often affects the final shape of the seeds. Plants generally produce ovules of four shapes: the most common shape is called anatropous, with a curved shape. Orthotropous ovules are straight with all the parts of the ovule lined up in a long row producing an uncurved seed. Campylotropous ovules have a curved megagametophyte often giving the seed a tight "C" shape. The last ovule shape is called amphitropous, where the ovule is partly inverted and turned back 90 degrees on its stalk (the funicle or funiculus).

In the majority of flowering plants, the zygote's first division is transversely oriented in regards to the long axis, and this establishes the polarity of the embryo. The upper or chalazal pole becomes the main area of growth of the embryo, while the lower or micropylar pole produces the stalk-like suspensor that attaches to the micropyle. The suspensor absorbs and manufacturers nutrients from the endosperm that are used during the embryo's growth.

Embryo

The main components of the embryo are:

- The cotyledons, the seed leaves, attached to the embryonic axis. There may be one (Mono-cotyledons), or two (Dicotyledons). The cotyledons are also the source of nutrients in the non-endospermic dicotyledons, in which case they replace the endosperm, and are thick and leathery. In endospermic seeds the cotyledons are thin and papery. Dicotyledons have the point of attachment opposite one another on the axis.

- The epicotyl, the embryonic axis above the point of attachment of the cotyledon(s).

- The plumule, the tip of the epicotyl, and has a feathery appearance due to the presence of young leaf primordia at the apex, and will become the shoot upon germination.

- The hypocotyl, the embryonic axis below the point of attachment of the cotyledon(s), connecting the epicotyl and the radicle, being the stem-root transition zone.

- The radicle, the basal tip of the hypocotyl, grows into the primary root.

The inside of a *Ginkgo* seed, showing a well-developed embryo,
nutritive tissue (megagametophyte), and a bit of the surrounding seed coat

Monocotyledonous plants have two additional structures in the form of sheaths. The plumule is covered with a coleoptile that forms the first leaf while the radicle is covered with a coleorhiza that connects to the primary root and adventitious roots form from the sides. Here the hypocotyl is a rudimentary axis between radicle and plumule. The seeds of corn are constructed with these structures; pericarp, scutellum (single large cotyledon) that absorbs nutrients from the endosperm, plumule, radicle, coleoptile and coleorhiza—these last two structures are sheath-like and enclose the plumule and radicle, acting as a protective covering.

Seed Coat

The maturing ovule undergoes marked changes in the integuments, generally a reduction and disorganisation but occasionally a thickening. The seed coat forms from the two integuments or outer layers of cells of the ovule, which derive from tissue from the mother plant, the inner integument forms the tegmen and the outer forms the testa. (The seed coats of some mononocotyledon plants, such as the grasses, are not distinct structures, but are fused with the fruit wall to form a pericarp.) The testae of both monocots and dicots are often marked with patterns and textured markings, or have wings or tufts of hair. When the seed coat forms from only one layer, it is also called the testa, though not all such testae are homologous from one species to the next. The funiculus abscises (detaches at fixed point – abscission zone), the scar forming an oval depression, the hilum. Anatropous ovules have a portion of the funiculus that is adnate (fused to the seed coat), and which forms a longitudinal ridge, or raphe, just above the hilum. In bitegmic ovules (e.g. *Gossypium* described here) both inner and outer integuments contribute to the seed coat formation. With continuing maturation the cells enlarge in the outer integument. While the inner epidermis may remain a single layer, it may also divide to produce two to three layers and accumulates starch, and is referred to as the colourless layer. By contrast the outer epidermis becomes tanniferous. The inner integument may consist of eight to fifteen layers. (Kozlowski 1972)

As the cells enlarge, and starch is deposited in the outer layers of the pigmented zone below the outer epidermis, this zone begins to lignify, while the cells of the outer epidermis enlarge radially and their walls thicken, with nucleus and cytoplasm compressed into the outer layer. these cells which are broader on their inner surface are called palisade cells. In the inner epidermis the cells also enlarge radially with plate like thickening of the walls. The mature inner integument has a palisade layer, a pigmented zone with 15-20 layers, while the innermost layer is known as the fringe layer. (Kozlowski 1972)

Gymnosperms

In gymnosperms, which do not form ovaries, the ovules and hence the seeds are exposed. This is the basis for their nomenclature – naked seeded plants. Two sperm cells transferred from the pollen do not develop the seed by double fertilization, but one sperm nucleus unites with the egg nucleus and the other sperm is not used.

Sometimes each sperm fertilizes an egg cell and one zygote is then aborted or absorbed during early development. The seed is composed of the embryo (the result of fertilization) and tissue from the mother plant, which also form a cone around the seed in coniferous plants such as pine and spruce.

Shape and Appearance

A large number of terms are used to describe seed shapes, many of which are largely self-explanatory such as *Bean-shaped* (reniform) – resembling a kidney, with lobed ends on either side of the hilum, *Square* or *Oblong* – angular with all sides more or less equal or longer than wide, *Triangular* – three sided, broadest below middle, *Elliptic* or *Ovate* or *Obovate* – rounded at both ends, or egg shaped (ovate or obovate, broader at one end), being rounded but either symmetrical about the middle or broader below the middle or broader above the middle.

Other less obvious terms include discoid (resembling a disc or plate, having both thickness and parallel faces and with a rounded margin), ellipsoid, globose (spherical), or subglobose (Inflated, but less than spherical), lenticular, oblong, ovoid, reniform and sectoroid. Striate seeds are striped with parallel, longitudinal lines or ridges. The commonest colours are brown and black, other colours are infrequent. The surface varies from highly polished to considerably roughened. The surface may have a variety of appendages. A seed coat with the consistency of cork is referred to as suberose. Other terms include crustaceous (hard, thin or brittle).

Structure

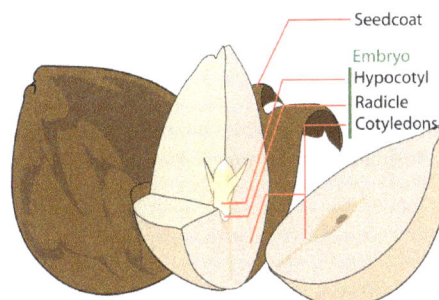

The parts of an avocado seed (a dicot), showing the seed coat and embryo

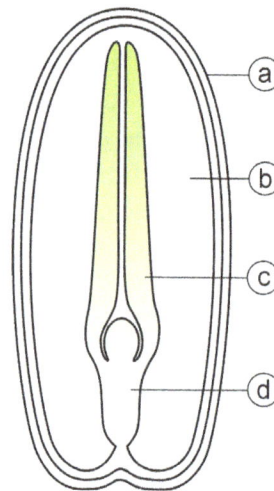

Diagram of the internal structure of a dicot seed and embryo: (a) seed coat, (b) endosperm, (c) cotyledon, (d) hypocotyl

A typical seed includes two basic parts:

1. an embryo;

2. a seed coat.

In addition, the endosperm forms a supply of nutrients for the embryo in most monocotyledons and the endospermic dicotyledons.

Seed Types

Seeds have been considered to occur in many structurally different types (Martin 1946). These are based on a number of criteria, of which the dominant one is the embryo-to-seed size ratio. This reflects the degree to which the developing cotyledons absorb the nutrients of the endosperm, and thus obliterate it.

Six types occur amongst the monocotyledons, ten in the dicotyledons, and two in the gymnosperms (linear and spatulate). This classification is based on three characteristics: embryo morphology, amount of endosperm and the position of the embryo relative to the endosperm.

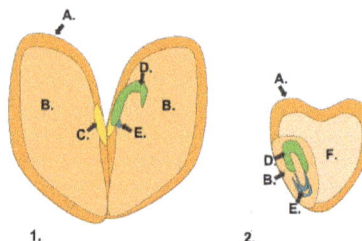

Diagram of a generalized dicot seed (1) versus a generalized monocot seed
(2). A. Seed Coat B. Cotyledon C. Hilum D. Plumule E. Radicle F. Endosperm

Embryo

In endospermic seeds, there are two distinct regions inside the seed coat, an upper and larger endosperm and a lower smaller embryo. The embryo is the fertilised ovule, an immature plant from which a new plant will grow under proper conditions. The embryo has one cotyledon or seed leaf in monocotyledons, two cotyledons in almost all dicotyledons and two or more in gymnosperms. In the fruit of grains (caryopses) the single monocotyledon is shield shaped and hence called a scutellum. The scutellum is pressed closely against the endosperm from which it absorbs food, and passes it to the growing parts. Embryo descriptors include small, straight, bent, curved and curled.

Nutrient Storage

Within the seed, there usually is a store of nutrients for the seedling that will grow from the embryo. The form of the stored nutrition varies depending on the kind of plant. In angiosperms, the stored food begins as a tissue called the endosperm, which is derived from the mother plant and the pollen via double fertilization. It is usually triploid, and is rich in oil or starch, and protein. In gymnosperms, such as conifers, the food storage tissue (also called endosperm) is part of the female gametophyte, a haploid tissue. The endosperm is surrounded by the aleurone layer (peripheral endosperm), filled with proteinaceous aleurone grains.

Originally, by analogy with the animal ovum, the outer nucellus layer (perisperm) was referred to as albumen, and the inner endosperm layer as vitellus. Although misleading, the term began to be applied to all the nutrient matter. This terminology persists in referring to endospermic seeds as "albuminous". The nature of this material is used in both describing and classifying seeds, in addition to the embryo to endosperm size ratio. The endosperm may be considered to be farinaceous (or mealy) in which the cells are filled with starch, as for instance cereal grains, or not (non-farinaceous). The endosperm may also be referred to as "fleshy" or "cartilaginous" with thicker soft cells such as coconut, but may also be oily as in *Ricinus* (castor oil), *Croton* and Poppy. The endosperm is called "horny" when the cell walls are thicker such as date and coffee, or "ruminated" if mottled, as in nutmeg, palms and Annonaceae.

In most monocotyledons (such as grasses and palms) and some (endospermic or albuminous) dicotyledons (such as castor beans) the embryo is embedded in the endosperm (and nucellus, which the seedling will use upon germination. In the non-endospermic dicotyledons the endosperm is absorbed by the embryo as the latter grows within the developing seed, and the cotyledons of the embryo become filled with stored food. At maturity, seeds of these species have no endosperm and are also referred to as exalbuminous seeds. The exalbuminous seeds include the legumes (such as beans and peas), trees such as the oak and walnut, vegetables such as squash and radish, and sunflowers. According to Bewley and Black (1978), Brazil nut storage is in hypocotyl, this place of storage is uncommon among seeds. All gymnosperm seeds are albuminous.

Seed Coat

The seed coat develops from the maternal tissue, the integuments, originally surrounding the ovule. The seed coat in the mature seed can be a paper-thin layer (e.g. peanut) or something more substantial (e.g. thick and hard in honey locust and coconut), or fleshy as in the sarcotesta of pomegranate. The seed coat helps protect the embryo from mechanical injury, predators and drying out. Depending on its development, the seed coat is either bitegmic or unitegmic. Bitegmic seeds form a testa from the outer integument and a tegmen from the inner integument while unitegmic seeds have only one integument. Usually parts of the testa or tegmen form a hard protective mechanical layer. The mechanical layer may prevent water penetration and germination. Amongst the barriers may be the presence of lignified sclereids.

The outer integument has a number of layers, generally between four and eight organised into three layers: (a) outer epidermis, (b) outer pigmented zone of two to five layers containing tannin and starch, and (c) inner epidermis. The endotegmen is derived from the inner epidermis of the inner integument, the exotegmen from the outer surface of the inner integument. The endotesta is derived from the inner epidermis of the outer integument, and the outer layer of the testa from the outer surface of the outer integument is referred to as the exotesta. If the exotesta is also the mechanical layer, this is called an exotestal seed, but if the mechanical layer is the endotegmen, then the seed is endotestal. The exotesta may consist of one or more rows of cells that are elongated and pallisade like (e.g. Fabaceae), hence 'palisade exotesta'.

In addition to the three basic seed parts, some seeds have an appendage, an aril, a fleshy outgrowth of the funicle (funiculus), (as in yew and nutmeg) or an oily appendage, an elaiosome (as in *Corydalis*), or hairs (trichomes). In the latter example these hairs are the source of the textile crop cotton. Other seed appendages include the raphe (a ridge), wings, caruncles (a soft spongy outgrowth from the outer integument in the vicinity of the micropyle), spines, or tubercles.

A scar also may remain on the seed coat, called the hilum, where the seed was attached to the ovary wall by the funicle. Just below it is a small pore, representing the micropyle of the ovule.

Size and Seed Set

A collection of various vegetable and herb seeds

Seeds are very diverse in size. The dust-like orchid seeds are the smallest, with about one million seeds per gram; they are often embryonic seeds with immature embryos and no significant energy reserves. Orchids and a few other groups of plants are mycoheterotrophs which depend on mycorrhizal fungi for nutrition during germination and the early growth of the seedling. Some terrestrial orchid seedlings, in fact, spend the first few years of their lives deriving energy from the fungi and do not produce green leaves. At over 20 kg, the largest seed is the *coco de mer*. Plants that produce smaller seeds can generate many more seeds per flower, while plants with larger seeds invest more resources into those seeds and normally produce fewer seeds. Small seeds are quicker to ripen and can be dispersed sooner, so fall blooming plants often have small seeds. Many annual plants produce great quantities of smaller seeds; this helps to ensure at least a few will end in a favorable place for growth. Herbaceous perennials and woody plants often have larger seeds; they can produce seeds over many years, and larger seeds have more energy reserves for germination and seedling growth and produce larger, more established seedlings after germination.

Functions

Seeds serve several functions for the plants that produce them. Key among these functions are nourishment of the embryo, dispersal to a new location, and dormancy during unfavorable conditions. Seeds fundamentally are means of reproduction, and most seeds are the product of sexual reproduction which produces a remixing of genetic material and phenotype variability on which natural selection acts.

Embryo Nourishment

Seeds protect and nourish the embryo or young plant. They usually give a seedling a faster start than a sporeling from a spore, because of the larger food reserves in the seed and the multicellularity of the enclosed embryo.

Dispersal

Unlike animals, plants are limited in their ability to seek out favorable conditions for life and growth. As a result, plants have evolved many ways to disperse their offspring by dispersing their seeds. A seed must somehow "arrive" at a location and be there at a time favorable for germination and growth. When the fruits open and release their seeds in a regular way, it is called dehiscent, which is often distinctive for related groups of plants; these fruits include capsules, follicles, legumes, silicles and siliques. When fruits do not open and release their seeds in a regular fashion, they are called indehiscent, which include the fruits achenes, caryopsis, nuts, samaras, and utricles.

By Wind (Anemochory)

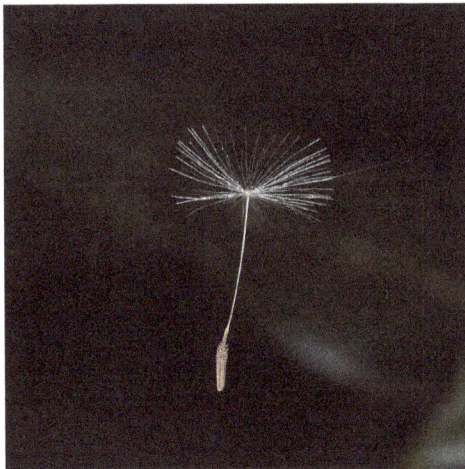

Dandelion seeds are contained within achenes, which can be carried long distances by the wind.

The seed pod of milkweed (*Asclepias syriaca*)

- Some seeds (e.g., pine) have a wing that aids in wind dispersal.

- The dustlike seeds of orchids are carried efficiently by the wind.

- Some seeds (e.g. milkweed, poplar) have hairs that aid in wind dispersal.

Other seeds are enclosed in fruit structures that aid wind dispersal in similar ways:

- Dandelion achenes have hairs.

- Maple samaras have two wings.

By Water (Hydrochory)

- Some plants, such as *Mucuna* and *Dioclea*, produce buoyant seeds termed sea-beans or drift seeds because they float in rivers to the oceans and wash up on beaches.

By Animals (Zoochory)

- Seeds (burrs) with barbs or hooks (e.g. acaena, burdock, dock) which attach to animal fur or feathers, and then drop off later.

- Seeds with a fleshy covering (e.g. apple, cherry, juniper) are eaten by animals (birds, mammals, reptiles, fish) which then disperse these seeds in their droppings.

- Seeds (nuts) are attractive long-term storable food resources for animals (e.g. acorns, hazelnut, walnut); the seeds are stored some distance from the parent plant, and some escape being eaten if the animal forgets them.

Myrmecochory is the dispersal of seeds by ants. Foraging ants disperse seeds which have appendages called elaiosomes (e.g. bloodroot, trilliums, acacias, and many species of Proteaceae). Elaiosomes are soft, fleshy structures that contain nutrients for animals that eat them. The ants carry such seeds back to their nest, where the elaiosomes are eaten. The remainder of the seed, which is hard and inedible to the ants, then germinates either within the nest or at a removal site where the seed has been discarded by the ants. This dispersal relationship is an example of mutualism, since the plants depend upon the ants to disperse seeds, while the ants depend upon the plants seeds for food. As a result, a drop in numbers of one partner can reduce success of the other. In South Africa, the Argentine ant (*Linepithema humile*) has invaded and displaced native species of ants. Unlike the native ant species, Argentine ants do not collect the seeds of *Mimetes cucullatus* or eat the elaiosomes. In areas where these ants have invaded, the numbers of *Mimetes* seedlings have dropped.

Dormancy

Seed dormancy has two main functions: the first is synchronizing germination with the optimal conditions for survival of the resulting seedling; the second is spreading germination of a batch of seeds over time so a catastrophe (e.g. late frosts, drought, herbivory) does not result in the death of all offspring of a plant (bet-hedging). Seed dormancy is defined as a seed failing to germinate under environmental conditions optimal for germination, normally when the environment is at a suitable temperature with proper soil moisture. This true dormancy or innate dormancy is therefore caused by conditions within the seed that prevent germination. Thus dormancy is a state of the seed, not of the environment. Induced dormancy, enforced dormancy or seed quiescence occurs when a seed fails to germinate because the external environmental conditions are inappropriate for germination, mostly in response to conditions being too dark or light, too cold or hot, or too dry.

Seed dormancy is not the same as seed persistence in the soil or on the plant, though even in scientific publications dormancy and persistence are often confused or used as synonyms.

Often, seed dormancy is divided into four major categories: exogenous; endogenous; combinational; and secondary. A more recent system distinguishes five classes: morphological, physiological, morphophysiological, physical and combinational dormancy.

Exogenous dormancy is caused by conditions outside the embryo, including:

- Physical dormancy or hard seed coats occurs when seeds are impermeable to water. At dormancy break, a specialized structure, the 'water gap', is disrupted in response to environmental cues, especially temperature, so water can enter the seed and germination can occur. Plant families where physical dormancy occurs include Anacardiaceae, Cannaceae, Convulvulaceae, Fabaceae and Malvaceae.

- Chemical dormancy considers species that lack physiological dormancy, but where a chemical prevents germination. This chemical can be leached out of the seed by rainwater or snow melt or be deactivated somehow. Leaching of chemical inhibitors from the seed by rain water is often cited as an important cause of dormancy release in seeds of desert plants, but little evidence exists to support this claim.

Endogenous dormancy is caused by conditions within the embryo itself, including:

- In morphological dormancy, germination is prevented due to morphological characteristics of the embryo. In some species, the embryo is just a mass of cells when seeds are dispersed; it is not differentiated. Before germination can take place, both differentiation and growth of the embryo have to occur. In other species, the embryo is differentiated but not fully grown (underdeveloped) at dispersal, and embryo growth up to a species specific length is required before germination can occur. Examples of plant families where morphological dormancy occurs are Apiaceae, Cycadaceae, Liliaceae, Magnoliaceae and Ranunculaceae.

- Morphophysiological dormancy includes seeds with underdeveloped embryos, and also have physiological components to dormancy. These seeds, therefore, require a dormancy-breaking treatments, as well as a period of time to develop fully grown embryos. Plant families where morphophysiological dormancy occurs include Apiaceae, Aquifoliaceae, Liliaceae, Magnoliaceae, Papaveraceae and Ranunculaceae. Some plants with morphophysiological dormancy, such as *Asarum* or *Trillium* species, have multiple types of dormancy, one affects radicle (root) growth, while the other affects plumule (shoot) growth. The terms "double dormancy" and "two-year seeds" are used for species whose seeds need two years to complete germination or at least two winters and one summer. Dormancy of the radicle (seedling root)is broken during the first winter after dispersal while dormancy of the shoot bud is broken during the second winter.

- Physiological dormancy means the embryo, due to physiological causes, cannot generate enough power to break through the seed coat, endosperm or other covering structures. Dormancy is typically broken at cool wet, warm wet or warm dry conditions. Abscisic acid is usually the growth inhibitor in seeds, and its production can be affected by light.

- o Drying, in some plants, including a number of grasses and those from seasonally arid regions, is needed before they will germinate. The seeds are released, but need to have a lower moisture content before germination can begin. If the seeds remain moist after dispersal, germination can be delayed for many months or even years. Many herbaceous plants from temperate climate zones have physiological dormancy that disappears with drying of the seeds. Other species will germinate after dispersal only under very narrow temperature ranges, but as the seeds dry, they are able to germinate over a wider temperature range.

- In seeds with combinational dormancy, the seed or fruit coat is impermeable to water and the embryo has physiological dormancy. Depending on the species, physical dormancy can be broken before or after physiological dormancy is broken.

- Secondary dormancy* is caused by conditions after the seed has been dispersed and occurs in some seeds when nondormant seed is exposed to conditions that are not favorable to germination, very often high temperatures. The mechanisms of secondary dormancy are not yet fully understood, but might involve the loss of sensitivity in receptors in the plasma membrane.

The following types of seed dormancy do not involve seed dormancy, strictly speaking, as lack of germination is prevented by the environment, not by characteristics of the seed itself:

- Photodormancy or light sensitivity affects germination of some seeds. These photoblastic seeds need a period of darkness or light to germinate. In species with thin seed coats, light may be able to penetrate into the dormant embryo. The presence of light or the absence of light may trigger the germination process, inhibiting germination in some seeds buried too deeply or in others not buried in the soil.

- Thermodormancy is seed sensitivity to heat or cold. Some seeds, including cocklebur and amaranth, germinate only at high temperatures (30 °C or 86 °F); many plants that have seeds that germinate in early to midsummer have thermodormancy, so germinate only when the soil temperature is warm. Other seeds need cool soils to germinate, while others, such as celery, are inhibited when soil temperatures are too warm. Often, thermodormancy requirements disappear as the seed ages or dries.

Not all seeds undergo a period of dormancy. Seeds of some mangroves are viviparous; they begin to germinate while still attached to the parent. The large, heavy root allows the seed to penetrate into the ground when it falls. Many garden plant seeds will germinate readily as soon as they have water and are warm enough; though their wild ancestors may have had dormancy, these cultivated plants lack it. After many generations of selective pressure by plant breeders and gardeners, dormancy has been selected out.

For annuals, seeds are a way for the species to survive dry or cold seasons. Ephemeral plants are usually annuals that can go from seed to seed in as few as six weeks.

Persistence and Seed Banks

Germination

Seed germination is a process by which a seed embryo develops into a seedling. It involves the reactivation of the metabolic pathways that lead to growth and the emergence of the radicle or seed root and plumule or shoot. The emergence of the seedling above the soil surface is the next phase of the plant's growth and is called seedling establishment.

Germinating sunflower seedlings

Three fundamental conditions must exist before germination can occur. (1) The embryo must be alive, called seed viability. (2) Any dormancy requirements that prevent germination must be overcome. (3) The proper environmental conditions must exist for germination.

Seed viability is the ability of the embryo to germinate and is affected by a number of different conditions. Some plants do not produce seeds that have functional complete embryos, or the seed may have no embryo at all, often called empty seeds. Predators and pathogens can damage or kill the seed while it is still in the fruit or after it is dispersed. Environmental conditions like flooding or heat can kill the seed before or during germination. The age of the seed affects its health and germination ability: since the seed has a living embryo, over time cells die and cannot be replaced. Some seeds can live for a long time before germination, while others can only survive for a short period after dispersal before they die.

Seed vigor is a measure of the quality of seed, and involves the viability of the seed, the germination percentage, germination rate and the strength of the seedlings produced.

The germination percentage is simply the proportion of seeds that germinate from all seeds subject to the right conditions for growth. The germination rate is the length of time it takes for the seeds to germinate. Germination percentages and rates are affected by seed viability, dormancy and environmental effects that impact on the seed and seedling. In agriculture and horticulture quality seeds have high viability, measured by germination percentage plus the rate of germination. This is given as a percent of germination over a certain amount of time, 90% germination in 20 days, for example. 'Dormancy' is covered above; many plants produce seeds with varying degrees of dormancy, and different seeds from the same fruit can have different degrees of dormancy. It's possible to have seeds with no dormancy if they are dispersed right away and do not dry (if the seeds dry they go into physiological dormancy). There is great variation amongst plants and a dormant seed is still a viable seed even though the germination rate might be very low.

Environmental conditions affecting seed germination include; water, oxygen, temperature and light.

Three distinct phases of seed germination occur: water imbibition; lag phase; and radicle emergence.

In order for the seed coat to split, the embryo must imbibe (soak up water), which causes it to swell, splitting the seed coat. However, the nature of the seed coat determines how rapidly water can penetrate and subsequently initiate germination. The rate of imbibition is dependent on the permeability of the seed coat, amount of water in the environment and the area of contact the seed has to the source of water. For some seeds, imbibing too much water too quickly can kill the seed. For some seeds, once water is imbibed the germination process cannot be stopped, and drying then becomes fatal. Other seeds can imbibe and lose water a few times without causing ill effects, but drying can cause secondary dormancy.

Repair of DNA Damage

During seed dormancy, often associated with unpredictable and stressful environments, DNA damages accumulate as the seeds age. In rye seeds, the reduction of DNA integrity due to damage is associated with loss of seed viability during storage. Upon germination, seeds of *Vicia faba* undergo DNA repair. A plant DNA ligase that is involved in repair of single- and double-strand breaks during seed germination is an important determinant of seed longevity. Also, in Arabidopsis seeds, the activities of the DNA repair enzymes Poly ADP ribose polymerases (PARP) are likely needed for successful germination. Thus DNA damages that accumulate during dormancy appear to be a problem for seed survival, and the enzymatic repair of DNA damages during germination appears to be important for seed viability.

Inducing Germination

A number of different strategies are used by gardeners and horticulturists to break seed dormancy.

Scarification allows water and gases to penetrate into the seed; it includes methods to physically break the hard seed coats or soften them by chemicals, such as soaking in hot water or poking holes in the seed with a pin or rubbing them on sandpaper or cracking with a press or hammer. Sometimes fruits are harvested while the seeds are still immature and the seed coat is not fully developed and sown right away before the seed coat become impermeable. Under natural conditions, seed coats are worn down by rodents chewing on the seed, the seeds rubbing against rocks (seeds are moved by the wind or water currents), by undergoing freezing and thawing of surface water, or passing through an animal's digestive tract. In the latter case, the seed coat protects the seed from digestion, while often weakening the seed coat such that the embryo is ready to sprout when it is deposited, along with a bit of fecal matter that acts as fertilizer, far from the parent plant. Microorganisms are often effective in breaking down hard seed coats and are sometimes used by people as a treatment; the seeds are stored in a moist warm sandy medium for several months under nonsterile conditions.

Stratification, also called moist-chilling, breaks down physiological dormancy, and involves the addition of moisture to the seeds so they absorb water, and they are then subjected to a period of moist chilling to after-ripen the embryo. Sowing in late summer and fall and allowing to overwinter under cool conditions is an effective way to stratify seeds; some seeds respond more favorably to periods of oscillating temperatures which are a part of the natural environment.

Leaching or the soaking in water removes chemical inhibitors in some seeds that prevent germination. Rain and melting snow naturally accomplish this task. For seeds planted in gardens, running water is best—if soaked in a container, 12 to 24 hours of soaking is sufficient. Soaking longer, especially in stagnant water, can result in oxygen starvation and seed death. Seeds with hard seed coats can be soaked in hot water to break open the impermeable cell layers that prevent water intake.

Other methods used to assist in the germination of seeds that have dormancy include prechilling, predrying, daily alternation of temperature, light exposure, potassium nitrate, the use of plant growth regulators, such as gibberellins, cytokinins, ethylene, thiourea, sodium hypochlorite, and others. Some seeds germinate best after a fire. For some seeds, fire cracks hard seed coats, while in others, chemical dormancy is broken in reaction to the presence of smoke. Liquid smoke is often used by gardeners to assist in the germination of these species.

Sterile Seeds

Seeds may be sterile for few reasons: they may have been irradiated, unpollinated, cells lived past expectancy, or bred for the purpose.

Evolution and Origin of Seeds

The origin of seed plants is a problem that still remains unsolved. However, more and more data tends to place this origin in the middle Devonian. The description in 2004 of the proto-seed *Runcaria heinzelinii* in the Givetian of Belgium is an indication of that ancient origin of seed-plants. As with modern ferns, most land plants before this time reproduced by sending spores into the air, that would land and become whole new plants.

The first "true" seeds are described from the upper Devonian, which is probably the theater of their true first evolutionary radiation. The seed plants progressively became one of the major elements of nearly all ecosystems.

Economic Importance

Phaseolus vulgaris seeds are diverse in size, shape, and color.

Edible Seeds

Many seeds are edible and the majority of human calories comes from seeds, especially from cereals, legumes and nuts. Seeds also provide most cooking oils, many beverages and spices and some important food additives. In different seeds the seed embryo or the endosperm dominates and provides most of the nutrients. The storage proteins of the embryo and endosperm differ in their amino acid content and physical properties. For example, the gluten of wheat, important in providing the elastic property to bread dough is strictly an endosperm protein.

Seeds are used to propagate many crops such as cereals, legumes, forest trees, turfgrasses and pasture grasses. Particularly in developing countries, a major constraint faced is the inadequacy of the marketing channels to get the seed to poor farmers. Thus the use of farmer-retained seed remains quite common.

Seeds are also eaten by animals, and are fed to livestock. Many seeds are used as birdseed.

Poison and Food Safety

While some seeds are edible, others are harmful, poisonous or deadly. Plants and seeds often contain chemical compounds to discourage herbivores and seed predators. In some cases, these compounds simply taste bad (such as in mustard), but other compounds are toxic or break down into toxic compounds within the digestive system. Children, being smaller than adults, are more susceptible to poisoning by plants and seeds.

A deadly poison, ricin, comes from seeds of the castor bean. Reported lethal doses are anywhere from two to eight seeds, though only a few deaths have been reported when castor beans have been ingested by animals.

In addition, seeds containing amygdalin—apple, apricot, bitter almond, peach, plum, cherry, quince, and others—when consumed in sufficient amounts, may cause cyanide poisoning. Other seeds that contain poisons include annona, cotton, custard apple, datura, uncooked durian, golden chain, horse-chestnut, larkspur, locoweed, lychee, nectarine, rambutan, rosary pea, sour sop, sugar apple, wisteria, and yew. The seeds of the strychnine tree are also poisonous, containing the poison strychnine.

The seeds of many legumes, including the common bean (*Phaseolus vulgaris*), contain proteins called lectins which can cause gastric distress if the beans are eaten without cooking. The common bean and many others, including the soybean, also contain trypsin inhibitors which interfere with the action of the digestive enzyme trypsin. Normal cooking processes degrade lectins and trypsin inhibitors to harmless forms.

Other uses

Cotton fiber grows attached to cotton plant seeds. Other seed fibers are from kapok and milkweed.

Many important nonfood oils are extracted from seeds. Linseed oil is used in paints. Oil from jojoba and crambe are similar to whale oil.

Seeds are the source of some medicines including castor oil, tea tree oil and the cancer drug, Laetrile.

Many seeds have been used as beads in necklaces and rosaries including Job's tears, Chinaberry, rosary pea, and castor bean. However, the latter three are also poisonous.

Other seed uses include:

- Seeds once used as weights for balances.

- Seeds used as toys by children, such as for the game Conkers.

- Resin from *Clusia rosea* seeds used to caulk boats.

- Nematicide from milkweed seeds.

- Cottonseed meal used as animal feed and fertilizer.

Seed Records

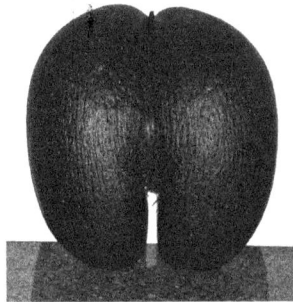

The massive fruit of the coco de mer

- The oldest viable carbon-14-dated seed that has grown into a plant was a Judean date palm seed about 2,000 years old, recovered from excavations at Herod the Great's palace on Masada in Israel. It was germinated in 2005. (A reported regeneration of *Silene stenophylla* (narrow-leafed campion) from material preserved for 31,800 years in the Siberian permafrost was achieved using fruit tissue, not seed.)

- The largest seed is produced by the coco de mer, or "double coconut palm", *Lodoicea maldivica*. The entire fruit may weigh up to 23 kilograms (50 pounds) and usually contains a single seed.

- The smallest seeds are produced by epiphytic orchids. They are only 85 micrometers long, and weigh 0.81 micrograms. They have no endosperm and contain underdeveloped embryos.

- The earliest fossil seeds are around 365 million years old from the Late Devonian of West Virginia. The seeds are preserved immature ovules of the plant *Elkinsia polymorpha*.

In Religion

The Book of Genesis in the Old Testament begins with an explanation of how all plant forms began:

And God said, Let the earth bring forth grass, the herb yielding seed, and the fruit tree yielding fruit after his kind, whose seed is in itself, upon the earth: and it was so. And the earth brought forth grass, and herb yielding seed after its kind, and the tree yielding fruit, whose seed was in itself, after its kind: and God saw that it was good. And the evening and the morning were the third day.

The Quran speaks about seed germination:

It is Allah Who causeth the seed-grain and the date-stone to split and sprout. He causeth the living to issue from the dead, and He is the one to cause the dead to issue from the living. That is Allah: then how are ye deluded away from the truth?

Ovule

Location of ovules inside a *Helleborus foetidus* flower

In seed plants, the ovule ("small egg") is the structure that gives rise to and contains the female reproductive cells. It consists of three parts: The integument(s) forming its outer layer(s), the nucellus (or remnant of the megasporangium), and female gametophyte (formed from haploid megaspore) in its center. The female gametophyte—specifically termed a *megagametophyte*—is also called the embryo sac in angiosperms. The megagametophyte produces an egg cell (or several egg cells in some groups) for the purpose of fertilization. After fertilization, the ovule develops into a seed.

Location within the Plant

In flowering plants, the ovule is located inside the portion of the flower called the gynoecium. The ovary of the gynoecium produces one or more ovules and ultimately becomes the fruit wall. Ovules are attached to the placenta in the ovary through a stalk-like structure known as a funiculus (plural, funiculi). Different patterns of ovule attachment, or placentation, can be found among plant species, these include:

- Apical placentation: The placenta is at the apex (top) of the ovary. Simple or compound ovary.

- Axile placentation: The ovary is divided into radial segments, with placentas in separate locules. Ventral sutures of carpels meet at the centre of the ovary. Placentae are along fused margins of carpels. Two or more carpels. (e.g. *Hibiscus, Citrus, Solanum*)

- Basal placentation: The placenta is at the base (bottom) of the ovary on a protrusion of the thalamus (receptacle). Simple or compound carpel, unilocular ovary. (e.g. *Sonchus, Helianthus*, Compositae)

- Free-central placentation: Derived from axile as partitions are absorbed, leaving ovules at the central axis. Compound unilocular ovary. (e.g. *Stellaria, Dianthus*)

- Marginal placentation: Simplest type. There is only one elongated placenta on one side of the ovary, as ovules are attached at the fusion line of the carpel's margins. This is conspicuous in legumes. Simple carpel, unilocular ovary. (e.g. *Pisum*)

- Parietal placentation: Placentae on inner ovary wall within a non-sectioned ovary, corresponding to fused carpel margins. Two or more carpels, unilocular ovary. (e.g. *Brassica*)

- Superficial: Similar to axile, but placentae are on inner surfaces of multilocular ovary (e.g. *Nymphaea*)

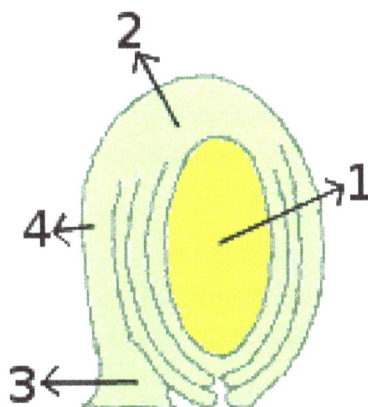

Ovule structure (anatropous) 1: nucleus 2: chalaza 3: funiculus 4: raphe

Ovule orientation may be anatropous, such that when inverted the micropyle faces the placenta (this is the most common ovule orientation in flowering plants), amphitropous, campylotropous, or orthotropous.

In gymnosperms such as conifers, ovules are borne on the surface of an ovuliferous (ovule-bearing) scale, usually within an ovulate cone (also called megastrobilus). In some extinct plants (e.g. Pteridosperms), megasporangia and perhaps ovules were borne on the surface of leaves. In other extinct taxa, a cupule (a modified leaf or part of a leaf) surrounds the ovule (e.g. *Caytonia* or *Glossopteris*).

Ovule Parts and Development

The ovule appears to be a megasporangium with integuments surrounding it. Ovules are initially composed of diploid maternal tissue, which includes a megasporocyte (a cell that will undergo meiosis to produce megaspores). Megaspores remain inside the ovule and divide by mitosis to produce the haploid female gametophyte or megagametophyte, which also remains inside the ovule. The remnants of the megasporangium tissue (the nucellus) surround the megagametophyte. Megagametophytes produce archegonia (lost in some groups such as flowering plants), which produce egg cells. After fertilization, the ovule contains a diploid zygote and then, after cell division begins, an embryo of the next sporophyte generation. In flowering plants, a second sperm nucleus fuses with other nuclei in the megagametophyte forming a typically polyploid (often triploid) endosperm tissue, which serves as nourishment for the young sporophyte.

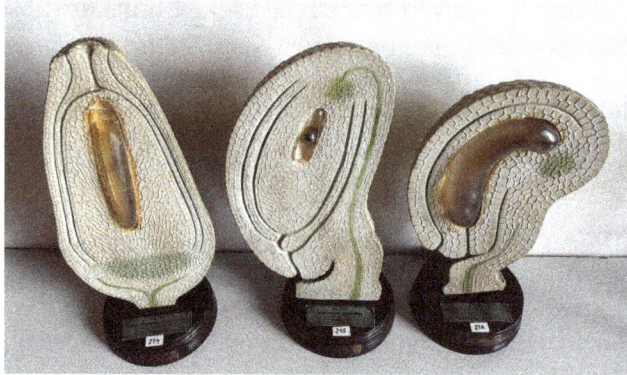

Models of different ovules, Botanical Museum Greifswald

Integuments, Micropyle and Chalaza

An integument is a protective cell layer surrounding the ovule. Gymnosperms typically have one integument (unitegmic) while angiosperms typically have two (bitegmic). The evolutionary origin of the inner integument (which is integral to the formation of ovules from megasporangia) has been proposed to be by enclosure of a megasporangium by sterile branches (telomes). *Elkinsia*, a preovulate taxon, has a lobed structure fused to the lower third of the megasporangium, with the lobes extending upwards in a ring around the megasporangium. This might, through fusion between lobes and between the structure and the megasporangium, have produced an integument.

The origin of the second or outer integument has been an area of active contention for some time. The cupules of some extinct taxa have been suggested as the origin of the outer integument. A few angiosperms produce vascular tissue in the outer integument, the orientation of which suggests that the outer surface is morphologically abaxial. This suggests that cupules of the kind produced by the Caytoniales or Glossopteridales may have evolved into the outer integument of angiosperms.

The integuments develop into the seed coat when the ovule matures after fertilization.

The integuments do not enclose the nucellus completely but retain an opening at the apex referred to as the micropyle. The micropyle opening allows the pollen (a male gametophyte) to enter the ovule for fertilization. In gymnosperms (e.g., conifers), the pollen is drawn into the ovule on a drop of fluid that exudes out of the micropyle, the so-called pollination drop mechanism. Subsequently, the micropyle closes. In angiosperms, only a pollen tube enters the micropyle. During germination, the seedling's radicle emerges through the micropyle.

Located opposite from the micropyle is the chalaza where the nucellus is joined to the integuments. Nutrients from the plant travel through the phloem of the vascular system to the funiculus and outer integument and from there apoplastically and symplastically through the chalaza to the nucellus inside the ovule. In chalazogamous plants, the pollen tubes enter the ovule through the chalaza instead of the micropyle opening.

Nucellus, Megaspore and Perisperm

The nucellus (plural: nucelli) is part of the inner structure of the ovule, forming a layer of diploid (sporophytic) cells immediately inside the integuments. It is structurally and functionally equivalent to the megasporangium. In immature ovules, the nucellus contains a megasporocyte (megaspore mother cell), which undergoes sporogenesis via meiosis. In gymnosperms, three of the four haploid spores produced in meiosis typically degenerate, leaving one surviving megaspore inside the nucellus. Among angiosperms, however, a wide range of variation exists in what happens next. The number (and position) of surviving megaspores, the total number of cell divisions, whether nuclear fusions occur, and the final number, position and ploidy of the cells or nuclei all vary. A common pattern of embryo sac development (the *Polygonum* type maturation pattern) includes a single functional megaspore followed by three rounds of mitosis. In some cases, however, two megaspores survive (for example, in *Allium* and *Endymion*). In some cases all four megaspores survive, for example in the *Fritillaria* type of development there is no separation of the megaspores following meiosis, then the nuclei fuse to form a triploid nucleus and a haploid nucleus. The subsequent arrangement of cells is similar to the *Polygonum* pattern, but the ploidy of the nuclei is different.

After fertilization, the nucellus may develop into the perisperm that feeds the embryo. In some plants, the diploid tissue of the nucellus can give rise to the embryo within the seed through a mechanism of asexual reproduction called nucellar embryony.

Megagametophyte

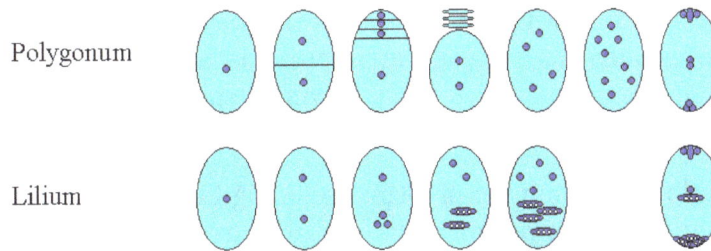

Megagametophyte formation of the genera *Polygonum* and *Lilium*. Triploid nuclei are shown as ellipses with three white dots. The first three columns show the meiosis of the megaspore, followed by 1-2 mitoses.

The haploid megaspore inside the nucellus gives rise to the female gametophyte, called the megagametophyte.

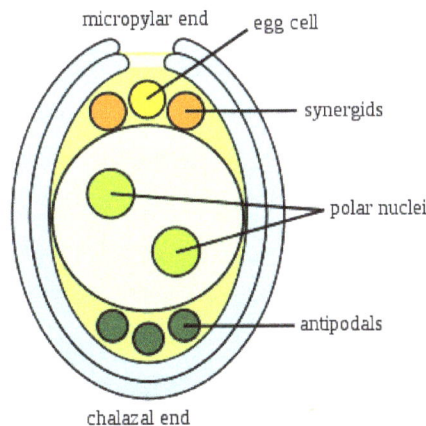

Ovule with megagametophyte: egg cell (yellow), synergids (orange),
central cell with two polar nuclei (bright green), and antipodals (dark green)

In gymnosperms, the megagametophyte consists of around 2000 nuclei and forms archegonia, which produce egg cells for fertilization.

In flowering plants, the megagametophyte (also referred to as the embryo sac) is much smaller and typically consists of only seven cells and eight nuclei. This type of megagametophyte develops from the megaspore through three rounds of mitotic divisions. The cell closest to the micropyle opening of the integuments differentiates into the egg cell, with two synergid cells by its side that are involved in the production of signals that guide the pollen tube. Three antipodal cells form on the opposite (chalazal) end of the ovule and later degenerate. The large central cell of the embryo sac contains two polar nuclei.

Zygote, Embryo and Endosperm

The pollen tube releases two sperm nuclei into the ovule. In gymnosperms, fertilization occurs within the archegonia produced by the female gametophyte. While it is possible that several egg cells are present and fertilized, typically only one zygote will develop into a mature embryo as the resources within the seed are limited.

In flowering plants, one sperm nucleus fuses with the egg cell to produce a zygote, the other fuses with the two polar nuclei of the central cell to give rise to the polyploid (typically triploid) endosperm. This double fertilization is unique to flowering plants, although in some other groups the second sperm cell does fuse with another cell in the megagametophyte to produce a second embryo. The plant stores nutrients such as starch, proteins, and oils in the endosperm as a food source for the developing embryo and seedling, serving a similar function to the yolk of animal eggs. The endosperm is also called the albumen of the seed.

Embryos may be described by a number of terms including Linear (embryos have axile placentation and are longer than broad), or rudimentary (embryos are basal in which the embryo is tiny in relation to the endosperm).

Types of Gametophytes

Megagametophytes of flowering plants may be described according to the number of megaspores developing, as either monosporic, bisporic, or tetrasporic.

Cotyledon

Cotyledon from a Judas-tree (*Cercis siliquastrum*) seedling

A cotyledon is a significant part of the embryo within the seed of a plant, and is defined by the Oxford English Dictionary as "The primary leaf in the embryo of the higher plants (Phanerogams); the seed-leaf." Upon germination, the cotyledon may become the embryonic first leaves of a seedling. The number of cotyledons present is one characteristic used by botanists to classify the flowering plants (angiosperms). Species with one cotyledon are called monocotyledonous ("monocots"). Plants with two embryonic leaves are termed dicotyledonous ("dicots") and placed in the class Magnoliopsida.

Comparison of a monocot and dicot sprouting. The visible part of the monocot plant (left) is actually the first true leaf produced from the meristem; the cotyledon itself remains within the seed

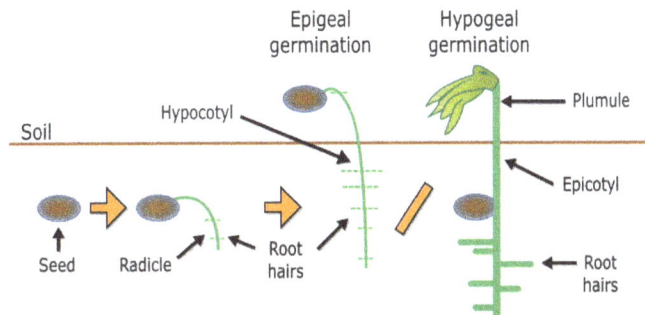

Schematic of epigeal vs hypogeal germination

In the case of dicot seedlings whose cotyledons are photosynthetic, the cotyledons are functionally similar to leaves. However, true leaves and cotyledons are developmentally distinct. Cotyledons are formed during embryogenesis, along with the root and shoot meristems, and are therefore present in the seed prior to germination. True leaves, however, are formed post-embryonically (i.e. after germination) from the shoot apical meristem, which is responsible for generating subsequent aerial portions of the plant.

The cotyledon of grasses and many other monocotyledons is a highly modified leaf composed of a *scutellum* and a *coleoptile*. The scutellum is a tissue within the seed that is specialized to absorb stored food from the adjacent endosperm. The coleoptile is a protective cap that covers the *plumule* (precursor to the stem and leaves of the plant).

Gymnosperm seedlings also have cotyledons, and these are often variable in number (multicotyledonous), with from 2 to 24 cotyledons forming a whorl at the top of the hypocotyl (the embryonic stem) surrounding the plumule. Within each species, there is often still some variation in cotyledon numbers, e.g. Monterey pine (*Pinus radiata*) seedlings have 5–9, and Jeffrey pine (*Pinus jeffreyi*) 7–13 (Mirov 1967), but other species are more fixed, with e.g. Mediterranean cypress always having just two cotyledons. The highest number reported is for big-cone pinyon (*Pinus maximartinezii*), with 24 (Farjon & Styles 1997).

The cotyledons may be ephemeral, lasting only days after emergence, or persistent, enduring at least a year on the plant. The cotyledons contain (or in the case of gymnosperms and monocotyledons, have access to) the stored food reserves of the seed. As these reserves are used up, the cotyledons may turn green and begin photosynthesis, or may wither as the first true leaves take over food production for the seedling.

Epigeal Versus Hypogeal Development

Cotyledons may be either epigeal, expanding on the germination of the seed, throwing off the seed shell, rising above the ground, and perhaps becoming photosynthetic; or hypogeal, not expanding, remaining below ground and not becoming photosynthetic. The latter is typically the case where the cotyledons act as a storage organ, as in many nuts and acorns.

Hypogeal plants have (on average) significantly larger seeds than epigeal ones. They are also capable of surviving if the seedling is clipped off, as meristem buds remain underground (with epigeal plants, the meristem is clipped off if the seedling is grazed). The tradeoff is whether the plant should produce a large number of small seeds, or a smaller number of seeds which are more likely to survive.

Related plants show a mixture of hypogeal and epigeal development, even within the same plant family. Groups which contain both hypogeal and epigeal species include, for example, the Araucariaceae family of Southern Hemisphere conifers, the Fabaceae (pea family), and the genus *Lilium*. The frequently garden grown common bean - Phaseolus vulgaris - is epigeal while the closely related runner bean - Phaseolus coccineus - is hypogeal.

History

The term *cotyledon* was coined by Marcello Malpighi (1628–1694). John Ray was the first botanist to recognize that some plants have two and others only one, and eventually the first to recognize the immense importance of this fact to systematics, in *Methodus plantarum* (1682).

Theophrastus (3rd or 4th century BC) and Albertus Magnus (13th century) may also have recognized the distinction between the dicotyledons and monocotyledons.

Bark

Bark is the outermost layers of stems and roots of woody plants. Plants with bark include trees, woody vines, and shrubs. Bark refers to all the tissues outside of the vascular cambium and is a nontechnical term. It overlays the wood and consists of the inner bark and the outer bark. The inner bark, which in older stems is living tissue, includes the innermost area of the periderm. The outer bark in older stems includes the dead tissue on the surface of the stems, along with parts of the innermost periderm and all the tissues on the outer side of the periderm. The outer bark on trees which lies external to the last formed periderm is also called the rhytidome.

Bark of mature *Mango* (*Mangifera indica*) showing lichen growth; Kolkata, India

Japanese Maple bark.

Products derived from bark include: bark shingle siding and wall coverings, spices and other flavorings, tanbark for tannin, resin, latex, medicines, poisons, various hallucinogenic chemicals and cork. Bark has been used to make cloth, canoes, and ropes and used as a surface for paintings and map making. A number of plants are also grown for their attractive or interesting bark colorations and surface textures or their bark is used as landscape mulch.

Botanic Description

Bark of *Leucadendron argenteum*

What is commonly called bark includes a number of different tissues. Cork is an external, secondary tissue that is impermeable to water and gases, and is also called the phellem. The cork is produced by the cork cambium which is a layer of meristematically active cells which serve as a lateral meristem for the periderm. The cork cambium, which is also called the phellogen, is normally only one cell layer thick and it divides periclinally to the outside producing cork. The phelloderm, which is not always present in all barks, is a layer of cells formed by and interior to the cork cambium. Together, the phellem (cork), phellogen (cork cambium) and phelloderm constitute the periderm.

Cork cell walls contain suberin, a waxy substance which protects the stem against water loss, the invasion of insects into the stem, and prevents infections by bacteria and fungal spores. The cambium tissues, i.e., the cork cambium and the vascular cambium, are the only parts of a woody stem where cell division occurs; undifferentiated cells in the vascular cambium divide rapidly to produce secondary xylem to the inside and secondary phloem to the outside. Phloem is a nutrient-conducting tissue composed of sieve tubes or sieve cells mixed with parenchyma and fibers. The cortex is the primary tissue of stems and roots. In stems the cortex is between the epidermis layer and the phloem, in roots the inner layer is not phloem but the pericycle.

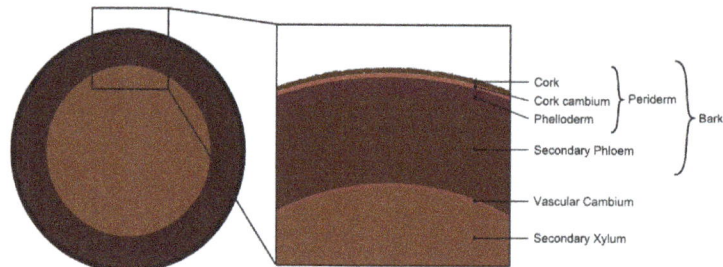

Tree cross section diagram

From the outside to the inside of a mature woody stem, the layers include:

1. Cork (Phellem)

2. Cork cambium (Phellogen)

3. Phelloderm

4. Cortex

5. Phloem

6. Vascular cambium

7. Xylem.

The bark includes (1) through (5), and is composed of periderm and phloem and the cells that produce these tissues. The periderms includes (1), (2) and (3).

Bark of a pine tree in Tecpán, Guatemala.

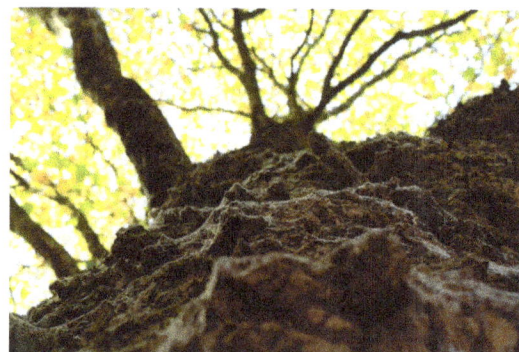

Bark of a tree in England

In young stems, which lack what is commonly called bark, the tissues are, from the outside to the inside: epidermis, periderm, cortex, primary phloem, secondary phloem, vascular cambium, secondary xylem, and primary xylem. As the stem ages and grows, changes occur that transform the surface of the stem into the bark. The epidermis is a layer of cells that cover the plant body, including the stems, leaves, flowers and fruits, that protects the plant from the outside world. In old stems the epidermal layer, cortex, and primary phloem become separated from the inner tissues by thicker formations of cork. Due to the thickening cork layer these cells die because they do not receive water and nutrients. This dead layer is the rough corky bark that forms around tree trunks and other stems.

Periderm

Often a secondary covering called the periderm forms on small woody stems and many non woody plants, which is composed of cork (phellem), the cork cambium (phellogen), and the phelloderm. The periderm forms from the phellogen which serves as a lateral meristem. The periderm replaces the epidermis, and acts as a protective covering like the epidermis. Mature phellem cells have suberin in their walls to protect the stem from desiccation and pathogen attack. Older phellem cells are dead, as is the case with woody stems. The skin on the potato tuber (which is an underground stem) constitutes the cork of the periderm.

In woody plants the epidermis of newly grown stems is replaced by the periderm later in the year. As the stems grow a layer of cells form under the epidermis, called the cork cambium, these cells produce cork cells that turn into cork. A limited number of cell layers may form interior to the cork cambium, called the phelloderm. As the stem grows, the cork cambium produces new layers of cork which are impermeable to gases and water and the cells outside of the periderm, namely the epidermis, cortex and older secondary phloem die.

Within the periderm are lenticels, which form during the production of the first periderm layer. Since there are living cells within the cambium layers that need to exchange gases during metabolism, these lenticels, because they have numerous intercellular spaces, allow gaseous exchange with the outside atmosphere. As the bark develops, new lenticels are formed within the cracks of the cork layers.

Rhytidome

Living tree bark enveloping barbed wire

The rhytidome is the most familiar part of bark, it is the outer layer that covers the trunks of trees. It is composed mostly of dead cells and is produced by the formation of multiple layers of suberized periderm, cortical and phloem tissue. It is generally thickest and most distinctive at the trunk or bole (the area from the ground to where the main branching starts) of the tree.

Chemical Composition

Bark tissues make up by weight between 10-20% of woody vascular plants and consists of various biopolymers, tannins, lignin, suberin, suberan and polysaccharides. Up to 40% of the bark tissue is made of lignin which forms an important part of a plant providing structural support by cross-linking between different polysaccharides, such as cellulose.

Condensed tannin, which are in fairly high concentration in bark tissue, is thought to inhibit decomposition. It could be due to this factor that the degradation of lignin is far less pronounced in bark tissue than it is in wood. It has been proposed that, in the cork layer (the phellogen), suberin acts as a barrier to microbial degradation and so protects the internal structure of the plant.

Analysis of the lignin in bark wall during decay by the white-rot fungi *Lentinula edodes* (Shiitake mushroom) using ^{13}C NMR revealed that the lignin polymers contained more Guaiacyl lignin units than Syringyl units compared to the interior of the plant. Guaiacyl units are less susceptible to degradation as, compared to syringyl, they contain fewer aryl-aryl bonds, can form a condensed lignin structure and have a lower redox potential. This could mean that the concentration and type of lignin units could provide additional resistance to fungal decay for plants protected by bark.

Uses

Bark chips

Cork, sometimes confused with bark in colloquial speech, is the outermost layer of a woody stem, derived from the cork cambium. It serves as protection against damage from parasites, herbivorous animals and diseases, as well as dehydration and fire. Cork can contain antiseptics like tannins, that protect against fungal and bacterial attacks that would cause decay.

Bark of pine was used as emergency food in Finland during famine, last time during and after civil war in 1918.

Backpack made of birch bark.

In some plants, the bark is substantially thicker, providing further protection and giving the bark a characteristically distinctive structure with deep ridges. In the cork oak (*Quercus suber*) the bark is thick enough to be harvested as a cork product without killing the tree; in this species the bark may get very thick (e.g. more than 20 cm has been reported). Some barks can be removed in long sheets; the smooth surfaced bark of birch trees has been used as a covering in the making of canoes, as the drainage layer in roofs, for shoes, backpacks etc. The most famous example of using birch bark for canoes is the birch canoes of North America.

The inner bark (phloem) of some trees is edible; in Scandinavia, bark bread is made from rye to which the toasted and ground innermost layer of bark of scots pine or birch is added. The Sami people of far northern Europe used large sheets of *Pinus sylvestris* bark that were removed in the spring, prepared and stored for use as a staple food resource and the inner bark was eaten fresh, dried or roasted.

Mechanical Bark Processing

Bark contains strong fibres known as bast, and there is a long tradition in northern Europe of using bark from coppiced young branches of the small-leaved lime (*Tilia cordata*) to produce cordage and rope, used for example in the rigging of Viking age longships.

Among the commercial products made from bark are cork, cinnamon, quinine (from the bark of Cinchona) and aspirin (from the bark of willow trees). The bark of some trees notably oak (*Quercus robur*) is a source of tannic acid, which is used in tanning. Bark chips generated as a by-product of lumber production are often used in bark mulch in western North America. Bark is important to the horticultural industry since in shredded form it is used for plants that do not thrive in ordinary soil, such as epiphytes.

Bark Chip Extraction

Wood Adhesives from Bark-Derived Phenols

> Wood Bark has lignin content and when it is pyrolyzed (subjected to high temperatures in the absence of oxygen), it yields a liquid bio-oil product rich in natural phenol derivatives. The phenol derivatives are isolated and recovered for application as a replacement for fossil-based phenols in phenol-formaldehyde (PF) resins used in Oriented Strand Board (OSB) and plywood.

Bark Removal

Cut logs are inflamed either just before cutting or before curing. Such logs and even trunks and branches found in their natural state of decay in forests, where the bark has fallen off, are said to be decorticated.

A number of living organisms live in or on bark, including insects, fungi and other plants like mosses, algae and other vascular plants. Many of these organisms are pathogens or parasites but some also have symbiotic relationships.

Bark Repair

The degree to which trees are able to repair gross physical damage to their bark is very variable. Some are able to produce a callus growth which heals over the wound rapidly, but leaves a clear scar, whilst others such as oaks do not produce an extensive callus repair.

Frost crack and sun scald are examples of damage found on tree bark which trees can repair to a degree, depending on the severity.

The patterns left in the bark of a Chinese Evergreen Elm after repeated visits by a Yellow-Bellied Sapsucker (woodpecker) in early 2012.

The self-repair of the Chinese Evergreen Elm showing new bark growth, lenticels, and other self-repair of the holes made by a Yellow-Bellied Sapsucker (woodpecker) about two years earlier.

Cortex (Botany)

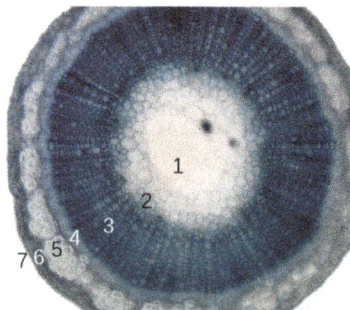

Cross-section of a flax plant stem:
1. Pith 2. Protoxylem 3. Xylem I 4. Phloem I 5. Sclerenchyma (bast fibre) 6. Cortex 7. Epidermis

A cortex is the outermost layer of a stem or root in a plant, or the surface layer or "skin" of the nonfruiting part of the body of some lichens.

In botany, the cortex is the outermost layer of the stem or root of a plant, bounded on the outside by the epidermis and on the inside by the endodermis. In plants, it is composed mostly of differentiated cells, usually large thin-walled parenchyma cells of the ground tissue system. The outer cortical cells often acquire irregularly thickened cell walls, and are called collenchyma cells. Some of the outer cortical cells may contain chloroplasts. It is responsible for the transportation of materials into the central cylinder of the root through diffusion and may also be used for food storage in the form of starch.

On a lichen, the cortex is the "skin", or outer layer of thallus tissue that covers the undifferentiated cells of the medulla. Fruticose lichens have one cortex encircling the branches, even flattened, leaf-like forms; foliose lichens have different upper and lower cortices; crustose, placodioid and squamulose lichens have an upper cortex but no lower cortex; and leprose lichens lack any cortex.

Cork Cambium

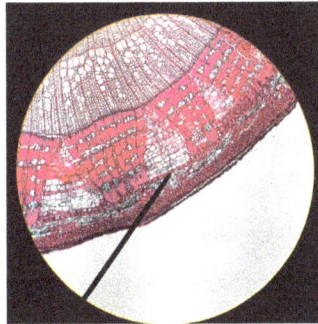

Cork cambium of woody stem (*Tilia*)

Cork cambium (pl. cambia or cambiums) is a tissue found in many vascular plants as part of the periderm. The cork cambium is a lateral meristem and is responsible for secondary growth that replaces the epidermis in roots and stems. It is found in woody and many herbaceous dicots, gymnosperms and some monocots (although monocots usually lack secondary growth). is one of the plant's meristems – the series of tissues consisting of embryonic (incompletely differentiated) cells from which the plant grows. It is one of the many layers of bark, between the cork and primary phloem. The function of cork cambium is to produce the cork, a tough protective material.

Synonyms for cork cambium are bark cambium, pericambium and phellogen. Phellogen is defined as the meristematic cell layer responsible for the development of the periderm. Cells that grow inwards from there are termed *phelloderm*, and cells that develop outwards are termed *phellem* or cork (similarity with vascular cambium). The periderm thus consists of three different layers:

- phelloderm – inside of cork cambium; composed of living parenchyma cells

- phellogen (cork cambium) – meristem that gives rise to periderm

- phellem (cork) – dead at maturity; air-filled protective tissue on the outside

Growth and development of cork cambium is very variable between different species, and is also highly dependent on age, growth conditions, etc. as can be observed from the different surfaces of bark: smooth, fissured, tesselated, scaly, flaking off, pie-like, etc.

Economic Importance

- Commercial cork is derived from the bark of the cork oak *(Quercus suber)*. Cork has many uses including wine bottle stoppers, bulletin boards, coasters, hot pads to protect tables from hot pans, insulation, sealing for lids, flooring, gaskets for engines, fishing bobbers, handles for fishing rods and tennis rackets, etc. It is also a high strength-to-weight/cost ablative material for aerodynamic prototypes in wind tunnels, as well as satellite launch vehicle payload fairings, reentry surfaces, and compression joints in thrust-vectored solid rocket motor nozzles.

- Many types of bark are used as mulch.

Phloem

Cross-section of a flax plant stem:

1. Pith

2. Protoxylem

3. MetaXylem I

4. Phloem I

5. Sclerenchyma (bast fibre)

6. Cortex

7. Epidermis

In vascular plants, phloem is the living tissue that transports the soluble organic compounds made during photosynthesis (known as photosynthate), in particular the sugar sucrose, to parts of the plant where needed. This transport process is called translocation. In trees, the phloem is the inner-most layer of the bark, hence the name, derived from the Greek word *(phloios)* meaning "bark".

Structure

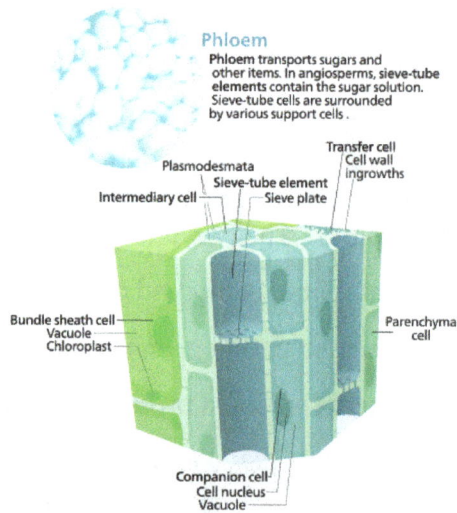

Cross section of some phloem cells

Phloem tissue consists of: conducting cells, generally called sieve elements; parenchyma cells, including both specialized companion cells or albuminous cells and unspecialized cells; and supportive cells, such as fibres and sclereids.

Conducting Cells (Sieve Elements)

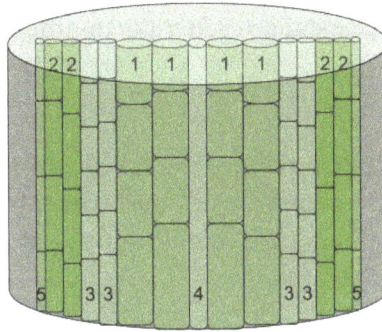

Simplified phloem and companion cells: 1.Xylem 2. Phloem 3. Cambium 4.Pith 5. Companion Cells

Sieve elements are the type of cell that are responsible for transporting sugars throughout the plant. At maturity they lack a nucleus and have very few organelles, so they rely on companion cells or albuminous cells for most of their metabolic needs. Sieve tube cells do contain vacuoles and other organelles, such as ribosomes, before they mature, but these generally migrate to the cell wall and dissolve at maturity; this ensures there is little to impede the movement of fluids. One of the few organelles they do contain at maturity is the smooth endoplasmic reticulum, which can be found at the plasma membrane, often nearby the plasmodesmata that connect them to their companion or albuminous cells. All sieve cells have groups of pores at their ends that grow from modified and enlarged plasmodesmata, called sieve areas. The pores are reinforced by platelets of a polysaccharide called callose.

Parenchyma Cells

Companion Cells

The metabolic functioning of sieve-tube members depends on a close association with the *companion cells*, a specialized form of parenchyma cell. All of the cellular functions of a sieve-tube element are carried out by the (much smaller) companion cell, a typical nucleate plant cell except the companion cell usually has a larger number of ribosomes and mitochondria. The dense cytoplasm of a companion cell is connected to the sieve-tube element by plasmodesmata. The common side-well shared by a sieve tube element and a companion cell has large numbers of plasmodesmata.

There are two types of companion cells.

1. Ordinary companion cells, which have smooth walls and few or no plasmodesmatal connections to cells other than the sieve tube.

2. Transfer cells, which have much-folded walls that are adjacent to non-sieve cells, allowing for larger areas of transfer. They are specialized in scavenging solutes from those in the cell walls that are actively pumped requiring energy.

Albuminous Cells

Albuminous cells have a similar role to companion cells, but are associated with sieve cells only and are hence found only in seedless vascular plants and gymnosperms.

Other Parenchyma Cells

Other parenchyma cells within the phloem are generally undifferentiated and used for food storage.

Supportive Cells

Although its primary function is transport of sugars, phloem may also contain cells that have a mechanical support function. These generally fall into two categories: fibres and sclereids. Both cell types have a secondary cell wall and are therefore dead at maturity. The secondary cell wall increases their rigidity and tensile strength.

Fibres

Bast fibres are the long, narrow supportive cells that provide tension strength without limiting flexibility. They are also found in xylem, and are the main component of many textiles such as paper, linen, and cotton.

Sclereids

Sclereids are irregularly shaped cells that add compression strength but may reduce flexibility to some extent. They also serve as anti-herbivory structures, as their irregular shape and hardness will increase wear on teeth as the herbivores chew. For example, they are responsible for the gritty texture in pears.

Function

Unlike xylem (which is composed primarily of dead cells), the phloem is composed of still-living cells that transport sap. The sap is a water-based solution, but rich in sugars made by photosynthesis. These sugars are transported to non-photosynthetic parts of the plant, such as the roots, or into storage structures, such as tubers or bulbs.

During the plant's growth period, usually during the spring, storage organs such as the roots are sugar sources, and the plant's many growing areas are sugar sinks. The movement in phloem is multidirectional, whereas, in xylem cells, it is unidirectional (upward).

After the growth period, when the meristems are dormant, the leaves are sources, and storage organs are sinks. Developing seed-bearing organs (such as fruit) are always sinks. Because of this multi-directional flow, coupled with the fact that sap cannot move with ease between adjacent sieve-tubes, it is not unusual for sap in adjacent sieve-tubes to be flowing in opposite directions.

While movement of water and minerals through the xylem is driven by negative pressures (tension) most of the time, movement through the phloem is driven by positive hydrostatic pressures. This process is termed *translocation*, and is accomplished by a process called phloem loading and *unloading*.

Phloem sap is also thought to play a role in sending informational signals throughout vascular plants. "Loading and unloading patterns are largely determined by the conductivity and number of plasmodesmata and the position-dependent function of solute-specific, plasma membrane transport proteins. Recent evidence indicates that mobile proteins and RNA are part of the plant's long-distance communication signaling system. Evidence also exists for the directed transport and sorting of macromolecules as they pass through plasmodesmata."

Organic molecules such as sugars, amino acids, certain hormones, and even messenger RNAs are transported in the phloem through sieve tube elements.

Girdling

Because phloem tubes are located outside the xylem in most plants, a tree or other plant can be killed by stripping away the bark in a ring on the trunk or stem. With the phloem destroyed, nutrients cannot reach the roots, and the tree/plant will die. Trees located in areas with animals such as beavers are vulnerable since beavers chew off the bark at a fairly precise height. This process is known as girdling, and can be used for agricultural purposes. For example, enormous fruits and vegetables seen at fairs and carnivals are produced via girdling. A farmer would place a girdle at the base of a large branch, and remove all but one fruit/vegetable from that branch. Thus, all the sugars manufactured by leaves on that branch have no sinks to go to but the one fruit/vegetable, which thus expands to many times normal size.

Origin

When the plant is an embryo, vascular tissue emerges from procambium tissue, which is at the center of the embryo. Protophloem itself appears in the mid-vein extending into the cotyledonary node, which constitutes the first appearance of a leaf in angiosperms, where it forms continuous strands. The hormone auxin, transported by the protein PIN1 is responsible for the growth of those protophloem strands, signaling the final identity of those tissues. SHORTROOT(SHR), and microRNA165/166 also participate in that process, while Callose Synthase 3(CALS3), inhibits the locations where SHORTROOT(SHR), and microRNA165 can go.

In the embryo, root phloem develops independently in the upper hypocotyl, which lies between the embryonic root, and the cotyledon.

In an adult, the phloem originates, and grows outwards from, meristematic cells in the vascular cambium. Phloem is produced in phases. *Primary* phloem is laid down by the apical meristem and develops from the procambium. *Secondary* phloem is laid down by the vascular cambium to the inside of the established layer(s) of phloem.

In some eudicot families (Apocynaceae, Convolvulaceae, Cucurbitaceae, Solanaceae, Myrtaceae, Asteraceae, Thymelaeaceae), phloem also develops on the inner side of the vascular cambium; in this case, a distinction between external phloem and internal phloem or intraxylary phloem is made. Internal phloem is mostly primary, and begins differentiation later than the external phloem and protoxylem, though it is not without exceptions. In some other families (Amaranthaceae, Nyctaginaceae, Salvadoraceae), the cambium also periodically forms inward strands or layers of phloem, embedded in the xylem: Such phloem strands are called included phloem or interxylary phloem.

Nutritional use

Phloem of pine trees has been used in Finland as a substitute food in times of famine and even in good years in the northeast. Supplies of phloem from previous years helped stave off starvation in the great famine of the 1860s. Phloem is dried and milled to flour (*pettu* in Finnish) and mixed with rye to form a hard dark bread, bark bread. The least appreciated was *silkko*, a bread made only from buttermilk and *pettu* without any real rye or cereal flour. Recently, *pettu* has again become available as a curiosity, and some have made claims of health benefits. However, its food energy content is low relative to rye or other cereals.

Phloem from silver birch has been also used to make flour in the past.

Vascular Cambium

The vascular cambium (also called main cambium, wood cambium, bifacial cambium; plural *cambia*) is a plant tissue located between the xylem and the phloem in the stems and roots of vascular plants. It is a cylinder of unspecialized meristem cells that divide to form secondary vascular tissues. It is the source of both secondary xylem growth inwards towards the pith, and secondary phloem growth outwards to the bark. Unlike the xylem and phloem, it does not transport water, minerals or food through the plant.

Vascular cambia are found in dicots and gymnosperms but not monocots, which usually lack secondary growth. A few leaf types also have a vascular cambium. In wood, the vascular cambium is the obvious line separating the bark and wood. For successful grafting, the vascular cambia of the rootstock and scion must be aligned so they can grow together.

Structure and Function

The cambium present between primary xylem and primary phloem is called *intrafasicular* cambium. During secondary growth, cells of meduallary rays, in a line with intrafasicular cambium, become meristematic and form *interfascicular* cambium. Therefore, the intrafascicular and interfascicular cambia form a ring which separates the primary xylem and primary phloem, and is known as *cambium ring*. The vascular cambium produces secondary xylem on the inside of the ring, and secondary phloem on the outside, pushing the primary xylem and phloem apart.

The vascular cambium usually consists of two types of cells:

- Fusiform initials (tall, axially oriented)
- Ray initials (smaller and round to angular in shape)

Maintenance of Cambial Meristem

The vascular cambium is maintained by a network of interacting signal feedback loops. Currently, both hormones and short peptides have been identified as information carriers in these systems. While similar regulation occurs in other meristems of plants, the cambial meristem receives signals from both the xylem and phloem sides for the meristem. Signals received from outside the meristem act to down regulate internal factors, which promotes cell proliferation, and promotes differentiation.

Hormonal Regulation

The phytohormones that are involved in the vascular cambial activity are auxins, ethylene, gibberellins, cytokinins, abscisic acid and more to be discovered. Each one of these plant hormones are vital for the regulation of the cambial activity and are dependent on their concentration.

Auxin hormones are proven to stimulate mitoses, cell production and regulate interfascicular and fascicular cambium. Applying auxin to the surface of a tree stump allowed decapitated shoots to continue secondary growth. The absence of auxin hormones will have a detrimental effect on a plant. It has been shown that mutants without auxin will exhibit increased spacing between the interfascicular cambiums and reduced growth of the vascular bundles. The mutant plant will therefore experience a decreased in water, nutrients, and photosynthates being transported throughout the plant, eventually leading to death. Auxin also regulates the two types of cell in the vascular cambium, ray and fusiform initials. Regulation of these initials ensures the connection and communication between xylem and phloem is maintained for the translocation of nourishment and sugars are safely being stored as an energy resource. Ethylene levels are high in plants with an active cambial zone and are still currently being studied. Gibberellin stimulates the cambial cell division and also regulates differentiation of the xylem tissues, with no effect on the rate of phloem differentiation. Differentiation is an essential process that changes these tissues into a more specialized type, leading to an important role in maintaining the life form of a plant. In poplar trees, high concentrations of gibberellin is positively correlated to an increase of cambial cell division and an increase of auxin in the cambial stem cells. Gibberellin is also responsible for the expansion of xylem through a signal traveling from the shoot to the root. Cytokinin hormone is known to regulate the rate of the cell division instead of the direction of cell differentiation. A study demonstrated that the mutants are found to have a reduction in stem and root growth but the secondary vascular pattern of the vascular bundles were not affected with a treatment of cytokinin.

Root

Primary and secondary roots in a cotton plant

In vascular plants, the root is the organ of a plant that typically lies below the surface of the soil. Roots can also be aerial or aerating, that is growing up above the ground or especially above water. Furthermore, a stem normally occurring below ground is not exceptional either. Therefore, the root is best defined as the non-leaf, non-nodes bearing parts of the plant's body. However, important internal structural differences between stems and roots exist.

Evolutionary History

The fossil record of roots – or rather, infilled voids where roots rotted after death – spans back to the late Silurian. Their identification is difficult, because casts and molds of roots are so similar in appearance to animal burrows. They can be discriminated using a range of features.

Definitions

The first root that comes from a plant is called the radicle. A root's four major functions are 1) absorption of water and inorganic nutrients, 2) anchoring of the plant body to the ground, and supporting it, 3) storage of food and nutrients, 4) vegetative reproduction and competition with other plants. In response to the concentration of nutrients, roots also synthesise cytokinin, which acts as a signal as to how fast the shoots can grow. Roots often function in storage of food and nutrients. The roots of most vascular plant species enter into symbiosis with certain fungi to form mycorrhizae, and a large range of other organisms including bacteria also closely associate with roots.

Large, mature tree roots above the soil

Anatomy

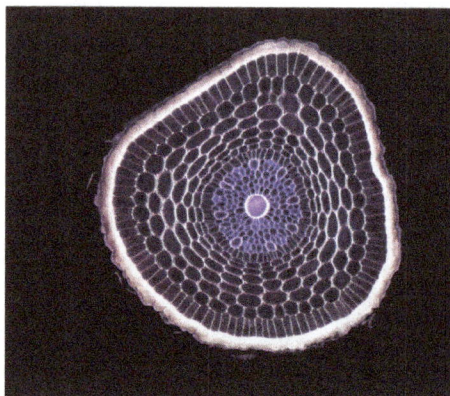

The cross-section of a barley root

When dissected, the arrangement of the cells in a root is root hair, epidermis, epiblem, cortex, endodermis, pericycle and, lastly, the vascular tissue in the centre of a root to transport the water absorbed by the root to other places of the plant.

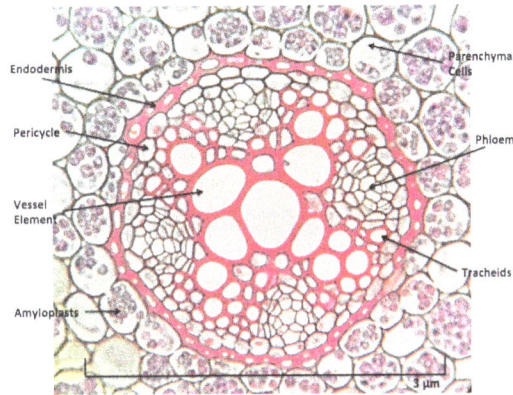

Ranunculus Root Cross Section

Architecture

Tree roots at Cliffs of the Neuse State Park

In its simplest form, the term root architecture refers to the spatial configuration of a plant's root system. This system can be extremely complex and is dependent upon multiple factors such as the species of the plant itself, the composition of the soil and the availability of nutrients.

The configuration of root systems serves to structurally support the plant, compete with other plants and for uptake of nutrients from the soil. Roots grow to specific conditions, which, if changed, can impede a plant's growth. For example, a root system that has developed in dry soil may not be as efficient in flooded soil, yet plants are able to adapt to other changes in the environment, such as seasonal changes.

Root architecture plays the important role of providing a secure supply of nutrients and water as well as anchorage and support. The main terms used to classify the architecture of a root system are:

- Branch magnitude: the number of links (exterior or interior).

- Topology: the pattern of branching, including:

1. Herringbone: alternate lateral branching off a parent root

2. Dichotomous: opposite, forked branches

3. Radial: whorl(s) of branches around a root

- Link length: the distance between branches.

- Root angle: the radial angle of a lateral root's base around the parent root's circumference, the angle of a lateral root from its parent root, and the angle an entire system spreads.

- Link radius: the diameter of a root.

All components of the root architecture are regulated through a complex interaction between genetic responses and responses due to environmental stimuli. These developmental stimuli are categorised as intrinsic, the genetic and nutritional influences, or extrinsic, the environmental influences and are interpreted by signal transduction pathways. The extrinsic factors that affect root architecture include gravity, light exposure, water and oxygen, as well as the availability or lack of nitrogen, phosphorus, sulphur, aluminium and sodium chloride. The main hormones (intrinsic stimuli) and respective pathways responsible for root architecture development include:

- Auxin – Auxin promotes root initiation, root emergence and primary root elongation.

- Cytokinins – Cytokinins regulate root apical meristem size and promote lateral root elongation.

- Gibberellins – Together with ethylene they promote crown primordia growth and elongation. Together with auxin they promote root elongation. Gibberellins also inhibit lateral root primordia initiation.

- Ethylene – Ethylene promotes crown root formation.

Growth

Roots of trees

Early root growth is one of the functions of the apical meristem located near the tip of the root. The meristem cells more or less continuously divide, producing more meristem, root cap cells (these are sacrificed to protect the meristem), and undifferentiated root cells. The latter become the primary tissues of the root, first undergoing elongation, a process that pushes the root tip forward in the growing medium. Gradually these cells differentiate and mature into specialized cells of the root tissues.

Growth from apical meristems is known as primary growth, which encompasses all elongation. Secondary growth encompasses all growth in diameter, a major component of woody plant tissues and many nonwoody plants. For example, storage roots of sweet potato have secondary growth but are not woody. Secondary growth occurs at the lateral meristems, namely the vascular cambium and cork cambium. The former forms secondary xylem and secondary phloem, while the latter forms the periderm.

In plants with secondary growth, the vascular cambium, originating between the xylem and the phloem, forms a cylinder of tissue along the stem and root. The vascular cambium forms new cells on both the inside and outside of the cambium cylinder, with those on the inside forming secondary xylem cells, and those on the outside forming secondary phloem cells. As secondary xylem accumulates, the "girth" (lateral dimensions) of the stem and root increases. As a result, tissues beyond the secondary phloem including the epidermis and cortex, in many cases tend to be pushed outward and are eventually "sloughed off" (shed).

At this point, the cork cambium begins to form the periderm, consisting of protective cork cells containing suberin. In roots, the cork cambium originates in the pericycle, a component of the vascular cylinder.

The vascular cambium produces new layers of secondary xylem annually. The xylem vessels are dead at maturity but are responsible for most water transport through the vascular tissue in stems and roots.

Tree roots usually grow to three times the diameter of the branch spread, only half of which lie underneath the trunk and canopy. The roots from one side of a tree usually supply nutrients to the foliage on the same side. Some families however, such as Sapindaceae (the maple family), show no correlation between root location and where the root supplies nutrients on the plant.

Regulation

There is a correlation of roots using the process of plant perception to sense their physical environment to grow, including the sensing of light, and physical barriers. Over time, roots can crack foundations, snap water lines, and lift sidewalks.

The correct environment of air, mineral nutrients and water directs plant roots to grow in any direction to meet the plant's needs. Roots will shy or shrink away from dry or other poor soil conditions.

Gravitropism directs roots to grow downward at germination, the growth mechanism of plants that also causes the shoot to grow upward.

Types

A true root system consists of a primary root and secondary roots (or lateral roots).

- the diffuse root system: the primary root is not dominant; the whole root system is fibrous and branches in all directions. Most common in monocots. The main function of the fibrous root is to anchor the plant.

Specialized

  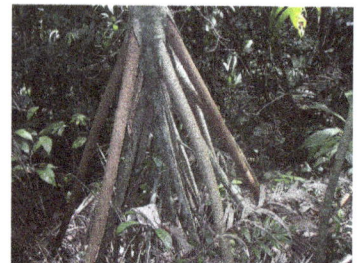

Aerial root   Stilt roots of Maize plant   The stilt roots of *Socratea exorrhiza*

The roots, or parts of roots, of many plant species have become specialized to serve adaptive purposes besides the two primary functions, described in the introduction.

- Adventitious roots arise out-of-sequence from the more usual root formation of branches of a primary root, and instead originate from the stem, branches, leaves, or old woody roots. They commonly occur in monocots and pteridophytes, but also in many dicots, such as clover (*Trifolium*), ivy (*Hedera*), strawberry (*Fragaria*) and willow (*Salix*). Most aerial roots and stilt roots are adventitious. In some conifers adventitious roots can form the largest part of the root system.

- Aerating roots (or knee root or knee or pneumatophores or Cypress knee): roots rising above the ground, especially above water such as in some mangrove genera (*Avicennia*, *Sonneratia*). In some plants like *Avicennia* the erect roots have a large number of breathing pores for exchange of gases.

- Aerial roots: roots entirely above the ground, such as in ivy (*Hedera*) or in epiphytic orchids. Many aerial roots, are used to receive water and nutrient intake directly from the air - from fogs, dew or humidity in the air. Some rely on leaf systems to gather rain or humidity and even store it in scales or pockets. Other aerial roots, such as mangrove aerial roots, are used for aeration and not for water absorption. Other aerial roots are used mainly for structure, functioning as prop roots, as in maize or anchor roots or as the trunk in strangler. In some Epiphytes - plants living above the surface on other plants, aerial roots serve for reaching to water sources or reaching the surface, and then functioning as regular surface roots.

- Contractile roots: they pull bulbs or corms of monocots, such as hyacinth and lily, and some taproots, such as dandelion, deeper in the soil through expanding radially and contracting longitudinally. They have a wrinkled surface.

- Coarse roots: Roots that have undergone secondary thickening and have a woody structure. These roots have some ability to absorb water and nutrients, but their main function is transport and to provide a structure to connect the smaller diameter, fine roots to the rest of the plant.

- Fine roots: Primary roots usually <2 mm diameter that have the function of water and nutrient uptake. They are often heavily branched and support mycorrhizas. These roots may be short lived, but are replaced by the plant in an ongoing process of root 'turnover'.

- Haustorial roots: roots of parasitic plants that can absorb water and nutrients from another plant, such as in mistletoe (*Viscum album*) and dodder.

- Propagative roots: roots that form adventitious buds that develop into aboveground shoots, termed suckers, which form new plants, as in Canada thistle, cherry and many others.

- Proteoid roots or cluster roots: dense clusters of rootlets of limited growth that develop under low phosphate or low iron conditions in Proteaceae and some plants from the following families Betulaceae, Casuarinaceae, Elaeagnaceae, Moraceae, Fabaceae and Myricaceae.

- Stilt roots: these are adventitious support roots, common among mangroves. They grow down from lateral branches, branching in the soil.

- Storage roots: these roots are modified for storage of food or water, such as carrots and beets. They include some taproots and tuberous roots.

- Structural roots: large roots that have undergone considerable secondary thickening and provide mechanical support to woody plants and trees.

- Surface roots: These proliferate close below the soil surface, exploiting water and easily available nutrients. Where conditions are close to optimum in the surface layers of soil, the growth of surface roots is encouraged and they commonly become the dominant roots.

- Tuberous roots: A portion of a root swells for food or water storage, e.g. sweet potato. A type of storage root distinct from taproot.

Depths

Cross section of a mango tree

The distribution of vascular plant roots within soil depends on plant form, the spatial and temporal availability of water and nutrients, and the physical properties of the soil. The deepest roots are generally found in deserts and temperate coniferous forests; the shallowest in tundra, boreal forest and temperate grasslands. The deepest observed living root, at least 60 metres below the ground surface, was observed during the excavation of an open-pit mine in Arizona, USA. Some roots can grow as deep as the tree is high. The majority of roots on most plants are however found relatively close to the surface where nutrient availability and aeration are more favourable for growth. Rooting depth may be physically restricted by rock or compacted soil close below the surface, or by anaerobic soil conditions.

Depth Records

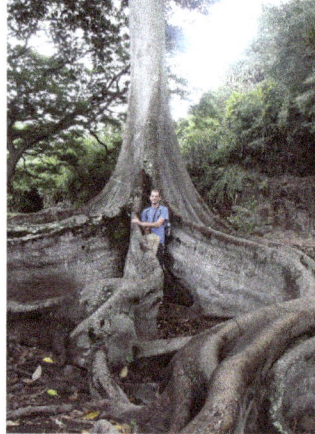

Ficus Tree with buttress roots

Environmental Interactions

Certain plants, namely Fabaceae, form root nodules in order to associate and form a symbiotic relationship with nitrogen-fixing bacteria called rhizobia. Due to the high energy required to fix nitrogen from the atmosphere, the bacteria take carbon compounds from the plant to fuel the process. In return, the plant takes nitrogen compounds produced from ammonia by the bacteria.

Economic Importance

Roots can also protect the environment by holding the soil to reduce soil erosion

The term root crops refers to any edible underground plant structure, but many root crops are actually stems, such as potato tubers. Edible roots include cassava, sweet potato, beet, carrot, rutabaga, turnip, parsnip, radish, yam and horseradish. Spices obtained from roots include sassafras, angelica, sarsaparilla and licorice.

Sugar beet is an important source of sugar. Yam roots are a source of estrogen compounds used in birth control pills. The fish poison and insecticide rotenone is obtained from roots of *Lonchocarpus* spp. Important medicines from roots are ginseng, aconite, ipecac, gentian and reserpine. Several legumes that have nitrogen-fixing root nodules are used as green manure crops, which provide nitrogen fertilizer for other crops when plowed under. Specialized bald cypress roots, termed knees, are sold as souvenirs, lamp bases and carved into folk art. Native Americans used the flexible roots of white spruce for basketry.

Tree roots can heave and destroy concrete sidewalks and crush or clog buried pipes. The aerial roots of strangler fig have damaged ancient Mayan temples in Central America and the temple of Angkor Wat in Cambodia.

Trees stabilize soil on a slope prone to landslides. The root hairs work as an anchor on the soil.

Vegetative propagation of plants via cuttings depends on adventitious root formation. Hundreds of millions of plants are propagated via cuttings annually including chrysanthemum, poinsettia, carnation, ornamental shrubs and many houseplants.

Roots can also protect the environment by holding the soil to reduce soil erosion. This is especially important in areas such as sand dunes.

Roots on onion bulbs

References

- Raven, Peter H.; Evert, Ray F.; Curtis, Helena (1981), Biology of Plants, New York, N.Y.: Worth Publishers, p. 641, ISBN 0-87901-132-7, OCLC 222047616

- Physics.org (2012). University of Adelaide. "Flightless parrots, burrowing bats helped parasitic Hades flower". Date Retrieved August 2013

- Soltis, Pamela S.; Douglas E. Soltis (2004). "The origin and diversification of angiosperms". American Journal of Botany. 91 (10): 1614–1626. PMID 21652312. doi:10.3732/ajb.91.10.1614

- Ausín, I.; et al. (2005). "Environmental regulation of flowering". Int J Dev Biol. 49 (5–6): 689–705. PMID 16096975. doi:10.1387/ijdb.052022ia

- Reynolds, Joan; Tampion, John (1983). Double flowers: a scientific study. London: [Published for the] Polytechnic of Central London Press [by] Pembridge Press. p. 41. ISBN 978-0-86206-004-6

- More, M, Cocucci, A.A, Raguso, R.A (2013). "The importance of oligosulfides in the attraction of fly pollinators to the brood-site deceptive species Jaborosa rotacea (Solanaceae)". International Journal of Plant Sciences. 174: 863–876. doi:10.1086/670367

- Igor Kosinki (2007). "Long-term variability in seed size and seedling establishment of Maianthemum bifolium". Plant Ecology. 194 (2): 149–156. doi:10.1007/s11258-007-9281-1

- Davis, P.H.; Cullen, J. (1979). The identification of flowering plant families, including a key to those native and cultivated in north temperate regions. Cambridge: Cambridge University Press. p. 106. ISBN 0-521-29359-6

- Sattler, R. & Lacroix, C. (1988). "Development and evolution of basal cauline placentation: Basella rubra". American Journal of Botany. 75: 918–927. doi:10.2307/2444012

- Beentje, Henk (2010). The Kew Plant Glossary. Richmond, Surrey: Royal Botanic Gardens, Kew. ISBN 978-1-84246-422-9. , p. 10

- Myking, T.; Hertzberg, A.; Skrøppa, T. (2005). "History, manufacture and properties of lime bast cordage in northern Europe". Forestry. 78 (1): 65–71. doi:10.1093/forestry/cpi006

- Macdonald, A.D. & Sattler, R. (1973). "Floral development of Myrica gale and the controversy over floral theories". Canadian Journal of Botany. 51: 1965–1975. doi:10.1139/b73-251

- Donald R. Whitehead (1969). "Wind Pollination in the Angiosperms: Evolutionary and Environmental Considerations". Evolution. 23 (1): 28–35. doi:10.2307/2406479

- Sattler, R. (1973). Organogenesis of Flowers : a Photographic Text-Atlas. University of Toronto Press. ISBN 978-0-8020-1864-9

- Vleeshouwers L.M., Bouwmeester H.J., Karssen C.M. (1995). "Redefining seed dormancy: an attempt to integrate physiology and ecology". Journal of Ecology. 83: 1031–1037. doi:10.2307/2261184

- Rost, Thomas L.; Weier, T. Elliot; Weier, Thomas Elliot (1979). Botany: a brief introduction to plant biology. New York: Wiley. p. 319. ISBN 0-471-02114-8

- L. Anders Nilsson (1988). "The evolution of flowers with deep corolla tubes". Nature. 334 (6178): 147–149. Bibcode:1988Natur.334..147N. doi:10.1038/334147a0

- The Seed Biology Place, Gerhard Leubner Lab, Royal Holloway, University of London, retrieved 13 October 2015

- Kolattukudy, P.E. (1984). "Biochemistry and function of cutin and suberin". Canadian Journal of Botany. 62: 2918–2933. doi:10.1139/b84-391

- Morhardt, Sia; Morhardt, Emil; Emil Morhardt, J. (2004). California desert flowers: an introduction to families, genera, and species. Berkeley: University of California Press. p. 24. ISBN 0-520-24003-0

- Thompson K, Ceriani RM, Bakker JP, Bekker RM (2003). "Are seed dormancy and persistence in soil related?". 13: 97–100. doi:10.1079/ssr2003128

- Ewers, F.W. (1982). "Secondary growth in needle leaves of Pinus longaeva (bristlecone pine) and other conifers: Quantitative data". American Journal of Botany. 69: 1552–1559. JSTOR 2442909. doi:10.2307/2442909

- Hartmann, Hudson Thomas, and Dale E. Kester. 1983. Plant propagation principles and practices. Englewood Cliffs, N.J.: Prentice-Hall. ISBN 0-13-681007-1. Pages 175-77

- Stewart, W.N.; Rothwell, G.W. (1993). Paleobotany and the evolution of plants. Cambridge University Press. ISBN 0521382947

- Peterson, R.L.; Barker, W.G. (1979). "Early tuber development from explanted stolon nodes of Solanum tuberosum var. Kennebec". Bot. Gaz. 140: 398–406. doi:10.1086/337104

Various Types of Tissues in Plants

Plant secretory tissues can be divided into lactiferous tissues and glandular tissues. Along with plant secretory tissues, the other tissues discussed are vascular tissues, spongy tissues and ground tissues. The aspects elucidated in this chapter are of vital importance, and provide a better understanding of plant anatomy.

Plant Secretory Tissue

The tissues that are concerned with the secretion of gums, resins, volatile oils, nectar latex, and other substances in plants are called secretory tissues These tissues are divided into two groups-

1. Laticiferous tissues

2. Glandular tissues

Introduction Class 11 and 12

Cells or organizations of cells which produce a variety of secretions. The secreted substance may remain deposited within the secretory cell itself or may be excreted, that is, released from the cell. Substances may be excreted to the surface of the plant or into intercellular cavities or canals. Some of the many substances contained in the secretions are not further utilized by the plant (resins, rubber, tannins, and various crystals), while others take part in the functions of the plant (enzymes and hormones). Secretory structures range from single cells scattered among other kinds of cells to complex structures involving many cells; the latter are often called glands.

Epidermal hairs of many plants are secretory or glandular. Such hairs commonly have a head composed of one or more secretory cells borne on a stalk. The hair of a stinging needle is bulbous below and extends into a long, fine process above. If one touches the hair, its tip breaks off, the sharp edge penetrates the skin, and the poisonous secretion is released. Glands secreting a sugary liquid—the nectar—in flowers pollinated by insects are called nectaries. Nectaries may occur on the floral stalk or on any floral organ: sepal, petal, stamen, or ovary.

The hydathode structures discharge water—a phenomenon called guttation through openings in margins or tips of leaves. The water flows through the xylem to its endings in the leaf and then through the intercellular spaces of the hydathode tissue toward the openings in the epidermis. Strictly speaking, such hydathodes are not glands because they are passive with regard to the flow of water.

Some carnivorous plants have glands that produce secretions capable of digesting insects and small animals. These glands occur on leaf parts modified as insect-trapping structures. In the sundews (Drosera) the traps bear stalked glands, called tentacles. When an insect lights on the leaf, the tentacles bend down and cover the victim with a mucilaginous secretion, the enzymes of which digest the insect.

Resin ducts are canals lined with secretory cells that release resins into the canal. Resin ducts are common in gymnosperms and occur in various tissues of roots, stems, leaves, and reproductive structures. Gum ducts are similar to resin ducts and may contain resins, oils, and gums. Usually, the term gum duct is used with reference to the dicotyledons, although gum ducts also may occur in the gymnosperms. Oil ducts are intercellular canals whose secretory cells produce oils or similar substances. Such ducts may be seen, for example, in various parts of the plant of the carrot family (Umbelliferae). Laticifers are cells or systems of cells containing latex, a milky or clear, colored or colorless liquid. Latex occurs under pressure and exudes from the plant when the latter is cut.

Laticiferous Tissues

These consist of thin walled, greatly elongated and much branched ducts containing a milky or yellowish colored juice known as latex. They contain numerous nuclei which lie embedded in the thin lining layer of protoplasm. They irregularly distributed in the mass of parenchymatous cells. Laticiferous ducts, in which latex are found are again two types-

1. Latex cell or non-articulate latex ducts

2. Latex vessels or articulate latex

Latex Cells

Also called as "non-articulate latex ducts", these ducts are independent units which extend as branched structures for long distances in the plant body. They originates as minute structures, elongate quickly and by repeated branching ramify in all directions but do not fuse together. Thus a network is not formed as in latex vessels.

Latex Vessel

Also called "articulate latex ducts", these ducts or vecssels are the result of anastamosing of many cells together. They grow more or less as parallel ducts which by means of branching and frequent anastamose form a complex network. Latex vessels are commonly found in many angiosperm families Papaveraceae, Compositae, Euphorbiaceae, Moraceae, etc.

Function

The function of laticiferous ducts is not clearly understood. They may also act as food storage organs or as reservoir of waste products. They may also act as translocatory tissues.

Glandular Tissues

This tissue consists of special structures; the glands. These glands contain some secretory or excretory products. A gland may consist of isolated cells or small group cells with or without a central cavity. They are of various kinds and may be internal or external. Internal glands are

• Oil-gland secreting essential oils, as in the fruits and leaves of orange, lemon.

- Mucilage secreting glands, as in the betel leaf

- Glands secreting gum, resin, tannin, etc.

- Digestive glands secreting enzymes or digestive agents

- Special water secreting glands at the tip of veins

External glands are commonly short hairs tipped by glands. They are:

- water-secreting hairs or glands,

- Glandular hairs secreting gum like substances as in tobacco, plumbago, etc.

- Glandular hairs secreting irritating, poisonous substances,as in nettles

- Honey glands, as in carnivorous plants.

Vascular Tissue

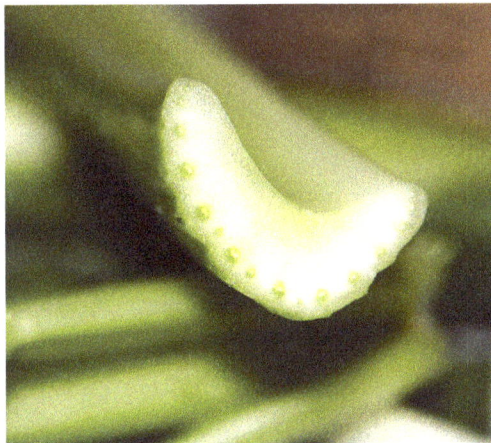

Cross section of celery stalk, showing vascular bundles, which include both phloem and xylem.

Vascular tissue is a complex conducting tissue, formed of more than one cell type, found in vascular plants. The primary components of vascular tissue are the xylem and phloem. These two tissues transport fluid and nutrients internally. There are also two meristems associated with vascular tissue: the vascular cambium and the cork cambium. All the vascular tissues within a particular plant together constitute the vascular tissue system of that plant.

The cells in vascular tissue are typically long and slender. Since the xylem and phloem function in the conduction of water, minerals, and nutrients throughout the plant, it is not surprising that their form should be similar to pipes. The individual cells of phloem are connected end-to-end, just as the sections of a pipe might be. As the plant grows, new vascular tissue differentiates in the growing tips of the plant. The new tissue is aligned with existing vascular tissue, maintaining its connection throughout the plant. The vascular tissue in plants is arranged in long, discrete strands called vascular bundles. These bundles include both xylem and phloem, as well as supporting and protective cells. In stems and roots, the xylem typically lies closer to the interior of the stem with phloem towards the exterior of the stem. In the stems of some Asterales dicots, there may be phloem located inwardly from the xylem as well.

Between the xylem and phloem is a meristem called the vascular cambium. This tissue divides off cells that will become additional xylem and phloem. This growth increases the girth of the plant, rather than its length. As long as the vascular cambium continues to produce new cells, the plant will continue to grow more stout. In trees and other plants that develop wood, the vascular cambium allows the expansion of vascular tissue that produces woody growth. Because this growth ruptures the epidermis of the stem, woody plants also have a cork cambium that develops among the phloem. The cork cambium gives rise to thickened cork cells to protect the surface of the plant and reduce water loss. Both the production of wood and the production of cork are forms of secondary growth.

In leaves, the vascular bundles are located among the spongy mesophyll. The xylem is oriented toward the adaxial surface of the leaf (usually the upper side), and phloem is oriented toward the abaxial surface of the leaf. This is why aphids are typically found on the underside of the leaves rather than on the top, since the phloem transports sugars manufactured by the plant and they are closer to the lower surface.

Xylem

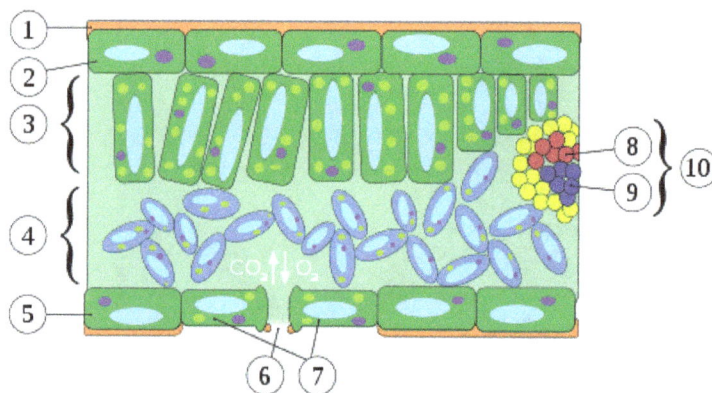

Schematic cross section of part of a leaf, xylem shown as red circles at 8

Xylem is one of the two types of transport tissue in vascular plants, phloem being the other. The basic function of xylem is to transport water from roots to shoots and leaves, but it also transports some nutrients. The word *xylem* is derived from the Greek word (*xylon*), meaning "wood"; the best-known xylem tissue is wood, though it is found throughout the plant.

Structure

The most distinctive xylem cells are the long tracheary elements that transport water. Tracheids and vessel elements are distinguished by their shape; vessel elements are shorter, and are connected together into long tubes that are called *vessels*.

Xylem also contains two other cell types: parenchyma and fibers.

Xylem can be found:

- in vascular bundles, present in non-woody plants and non-woody parts of woody plants

- in secondary xylem, laid down by a meristem called the vascular cambium in woody plants

- as part of a stelar arrangement not divided into bundles, as in many ferns.

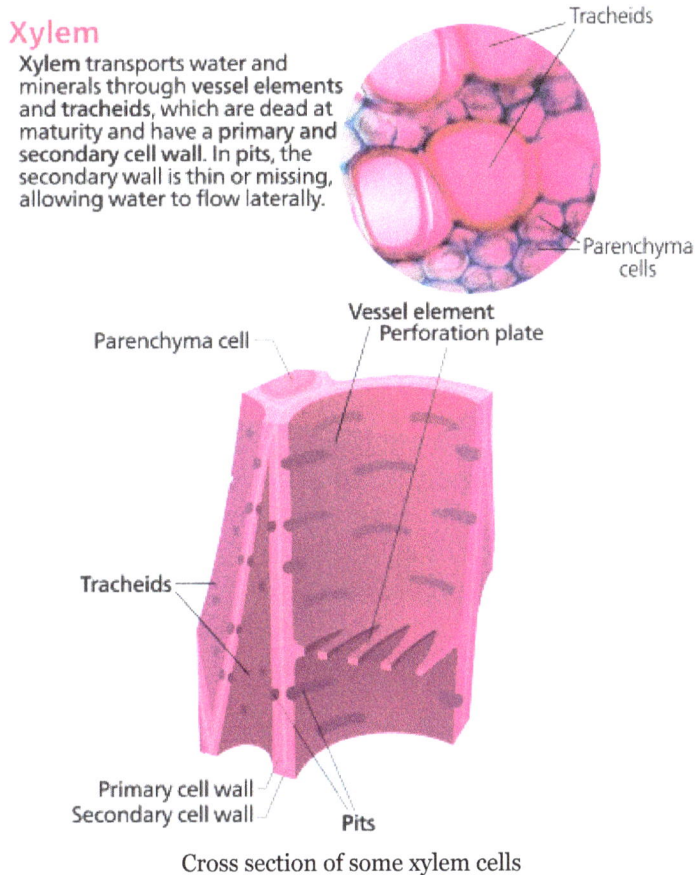

Cross section of some xylem cells

In transitional stages of plants with secondary growth, the first two categories are not mutually exclusive, although usually a vascular bundle will contain *primary xylem* only.

The branching pattern exhibited by xylem follows Murray's law.

Primary and Secondary Xylem

Primary xylem is formed during primary growth from procambium. It includes protoxylem and metaxylem. Metaxylem develops after the protoxylem but before secondary xylem. Metaxylem has wider vessels and tracheids than protoxylem.

Secondary xylem is formed during secondary growth from vascular cambium. Although secondary xylem is also found in members of the gymnosperm groups Gnetophyta and Ginkgophyta and to a lesser extent in members of the Cycadophyta, the two main groups in which secondary xylem can be found are:

1.   conifers (*Coniferae*): there are some six hundred species of conifers. All species have secondary xylem, which is relatively uniform in structure throughout this group. Many conifers become tall trees: the secondary xylem of such trees is used and marketed as softwood.

2. angiosperms (*Angiospermae*): there are some quarter of a million to four hundred thousand species of angiosperms. Within this group secondary xylem is rare in the monocots. Many non-monocot angiosperms become trees, and the secondary xylem of these is used and marketed as hardwood.

Main Function – Upwards Water Transport

The xylem transports water and soluble mineral nutrients from the roots throughout the plant. It is also used to replace water lost during transpiration and photosynthesis. Xylem sap consists mainly of water and inorganic ions, although it can contain a number of organic chemicals as well. The transport is passive, not powered by energy spent by the tracheary elements themselves, which are dead by maturity and no longer have living contents. Transporting sap upwards becomes more difficult as the height of a plant increases and upwards transport of water by xylem is considered to limit the maximum height of trees. Three phenomena cause xylem sap to flow:

- Pressure flow hypothesis: Sugars produced in the leaves and other green tissues are kept in the phloem system, creating a solute pressure differential versus the xylem system carrying a far lower load of solutes- water and minerals. The phloem pressure can rise to several MPa, far higher than atmospheric pressure. Selective inter-connection between these systems allows this high solute concentration in the phloem to draw xylem fluid upwards by negative pressure.

- Transpirational pull: Similarly, the evaporation of water from the surfaces of mesophyll cells to the atmosphere also creates a negative pressure at the top of a plant. This causes millions of minute menisci to form in the mesophyll cell wall. The resulting surface tension causes a negative pressure or tension in the xylem that pulls the water from the roots and soil.

- Root pressure: If the water potential of the root cells is more negative than that of the soil, usually due to high concentrations of solute, water can move by osmosis into the root from the soil. This causes a positive pressure that forces sap up the xylem towards the leaves. In some circumstances, the sap will be forced from the leaf through a hydathode in a phenomenon known as guttation. Root pressure is highest in the morning before the stomata open and allow transpiration to begin. Different plant species can have different root pressures even in a similar environment; examples include up to 145 kPa in *Vitis riparia* but around zero in *Celastrus orbiculatus*.

The primary force that creates the capillary action movement of water upwards in plants is the adhesion between the water and the surface of the xylem conduits. Capillary action provides the force that establishes an equilibrium configuration, balancing gravity. When transpiration removes water at the top, the flow is needed to return to the equilibrium.

Transpirational pull results from the evaporation of water from the surfaces of cells in the leaves. This evaporation causes the surface of the water to recess into the pores of the cell wall. By capillary action, the water forms concave menisci inside the pores. The high surface tension of water pulls the concavity outwards, generating enough force to lift water as high as a hundred meters from ground level to a tree's highest branches.

Transpirational pull requires that the vessels transporting the water are very small in diameter, otherwise cavitation would break the water column. And as water evaporates from leaves, more is drawn up through the plant to replace it. When the water pressure within the xylem reaches extreme levels due to low water input from the roots (if, for example, the soil is dry), then the gases come out of solution and form a bubble – an embolism forms, which will spread quickly to other adjacent cells, unless bordered pits are present (these have a plug-like structure called a torus, that seals off the opening between adjacent cells and stops the embolism from spreading).

Cohesion-tension Theory

The *cohesion-tension theory* is a theory of intermolecular attraction that explains the process of water flow upwards (against the force of gravity) through the xylem of plants. It was proposed in 1894 by John Joly and Henry Horatio Dixon. Despite numerous objections, this is the most widely accepted theory for the transport of water through a plant's vascular system based on the classical research of Dixon-Joly (1894), Askenasy (1895), and Dixon (1914,1924).

Water is a polar molecule. When two water molecules approach one another, the slightly negatively charged oxygen atom of one forms a hydrogen bond with a slightly positively charged hydrogen atom in the other. This attractive force, along with other intermolecular forces, is one of the principal factors responsible for the occurrence of surface tension in liquid water. It also allows plants to draw water from the root through the xylem to the leaf.

Water is constantly lost through transpiration from the leaf. When one water molecule is lost another is pulled along by the processes of cohesion and tension. Transpiration pull, utilizing capillary action and the inherent surface tension of water, is the primary mechanism of water movement in plants. However, it is not the only mechanism involved. Any use of water in leaves forces water to move into them.

Transpiration in leaves creates tension (differential pressure) in the cell walls of mesophyll cells. Because of this tension, water is being pulled up from the roots into the leaves, helped by cohesion (the pull between individual water molecules, due to hydrogen bonds) and adhesion (the stickiness between water molecules and the hydrophilic cell walls of plants). This mechanism of water flow works because of water potential (water flows from high to low potential), and the rules of simple diffusion.

Over the past century, there has been a great deal of research regarding the mechanism of xylem sap transport; today, most plant scientists continue to agree that the *cohesion-tension theory* best explains this process, but multiforce theories that hypothesize several alternative mechanisms have been suggested, including longitudinal cellular and xylem osmotic pressure gradients, axial potential gradients in the vessels, and gel- and gas-bubble-supported interfacial gradients.

Measurement of Pressure

Until recently, the differential pressure (suction) of transpirational pull could only be measured indirectly, by applying external pressure with a pressure bomb to counteract it. When the technology to perform direct measurements with a pressure probe was developed, there was initially some doubt about whether the classic theory was correct, because some workers were unable to demonstrate negative pressures. More recent measurements do tend to validate the classic theory, for the most part. Xylem transport is driven by a combination of transpirational pull from above and root pressure from below, which makes the interpretation of measurements more complicated.

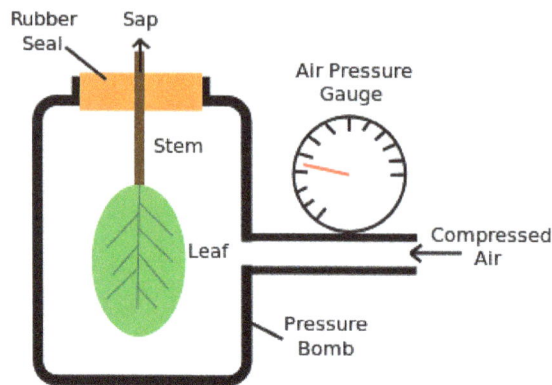

A diagram showing the setup of a pressure bomb

Evolution

Xylem appeared early in the history of terrestrial plant life. Fossil plants with anatomically preserved xylem are known from the Silurian (more than 400 million years ago), and trace fossils resembling individual xylem cells may be found in earlier Ordovician rocks. The earliest true and recognizable xylem consists of tracheids with a helical-annular reinforcing layer added to the cell wall. This is the only type of xylem found in the earliest vascular plants, and this type of cell continues to be found in the *protoxylem* (first-formed xylem) of all living groups of plants. Several groups of plants later developed pitted tracheid cells, it seems, through convergent evolution. In living plants, pitted tracheids do not appear in development until the maturation of the *metaxylem* (following the *protoxylem*).

In most plants, pitted tracheids function as the primary transport cells. The other type of tracheary element, besides the tracheid, is the vessel element. Vessel elements are joined by perforations into vessels. In vessels, water travels by *bulk flow*, as in a pipe, rather than by diffusion through cell membranes. The presence of vessels in xylem has been considered to be one of the key innovations that led to the success of the angiosperms. However, the occurrence of vessel elements is not restricted to angiosperms, and they are absent in some archaic or "basal" lineages of the angiosperms: (e.g., Amborellaceae, Tetracentraceae, Trochodendraceae, and Winteraceae), and their secondary xylem is described by Arthur Cronquist as "primitively vesselless". Cronquist considered the vessels of *Gnetum* to be convergent with those of angiosperms. Whether the absence of vessels in basal angiosperms is a primitive condition is contested, the alternative hypothesis states that vessel elements originated in a precursor to the angiosperms and were subsequently lost.

To photosynthesize, plants must absorb CO_2 from the atmosphere. However, this comes at a price: while stomata are open to allow CO_2 to enter, water can evaporate. Water is lost much faster than CO_2 is absorbed, so plants need to replace it, and have developed systems to transport water from the moist soil to the site of photosynthesis. Early plants sucked water between the walls of their cells, then evolved the ability to control water loss (and CO_2 acquisition) through the use of stomata. Specialized water transport tissues soon evolved in the form of hydroids, tracheids, then secondary xylem, followed by an endodermis and ultimately vessels.

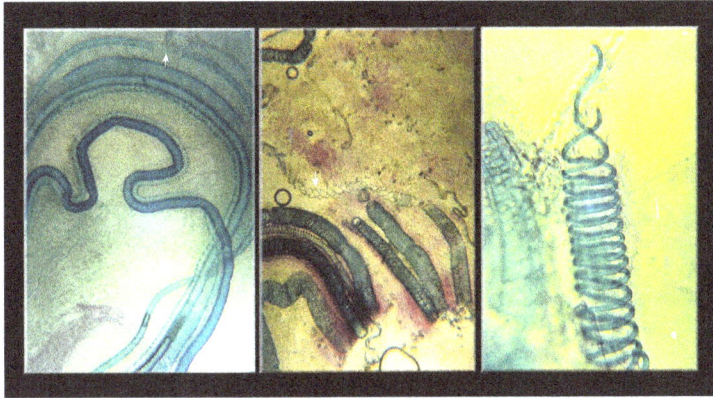

Photos showing xylem elements in the shoot of a fig tree (*Ficus alba*): crushed in hydrochloric acid, between slides and cover slips.

The high CO_2 levels of Silurian-Devonian times, when plants were first colonizing land, meant that the need for water was relatively low. As CO_2 was withdrawn from the atmosphere by plants, more water was lost in its capture, and more elegant transport mechanisms evolved. As water transport mechanisms, and waterproof cuticles, evolved, plants could survive without being continually covered by a film of water. This transition from poikilohydry to homoiohydry opened up new potential for colonization. Plants then needed a robust internal structure that held long narrow channels for transporting water from the soil to all the different parts of the above-soil plant, especially to the parts where photosynthesis occurred.

During the Silurian, CO_2 was readily available, so little water needed expending to acquire it. By the end of the Carboniferous, when CO_2 levels had lowered to something approaching today's, around 17 times more water was lost per unit of CO_2 uptake. However, even in these "easy" early days, water was at a premium, and had to be transported to parts of the plant from the wet soil to avoid desiccation. This early water transport took advantage of the cohesion-tension mechanism inherent in water. Water has a tendency to diffuse to areas that are drier, and this process is accelerated when water can be wicked along a fabric with small spaces. In small passages, such as that between the plant cell walls (or in tracheids), a column of water behaves like rubber – when molecules evaporate from one end, they pull the molecules behind them along the channels. Therefore, transpiration alone provided the driving force for water transport in early plants. However, without dedicated transport vessels, the cohesion-tension mechanism cannot transport water more than about 2 cm, severely limiting the size of the earliest plants. This process demands a steady supply of water from one end, to maintain the chains; to avoid exhausting it, plants developed a waterproof cuticle. Early cuticle may not have had pores but did not cover the entire plant surface, so that gas exchange could continue. However, dehydration at times was inevitable; early plants cope with this by having a lot of water stored between their cell walls, and when it comes to it sticking out the tough times by putting life "on hold" until more water is supplied.

To be free from the constraints of small size and constant moisture that the parenchymatic transport system inflicted, plants needed a more efficient water transport system. During the early Silurian, they developed specialized cells, which were lignified (or bore similar chemical compounds) to avoid implosion; this process coincided with cell death, allowing their innards to be emptied and water to be passed through them. These wider, dead, empty cells were a million times more conductive than the inter-cell method, giving the potential for transport over longer distances, and higher CO_2 diffusion rates.

A banded tube from the late Silurian/early Devonian. The bands are difficult to see on this specimen, as an opaque carbonaceous coating conceals much of the tube. Scale bar: 20 μm.

The earliest macrofossils to bear water-transport tubes are Silurian plants placed in the genus *Cooksonia*. The early Devonian pretracheophytes *Aglaophyton* and *Horneophyton* have structures very similar to the hydroids of modern mosses. Plants continued to innovate new ways of reducing the resistance to flow within their cells, thereby increasing the efficiency of their water transport. Bands on the walls of tubes, in fact apparent from the early Silurian onwards, are an early improvisation to aid the easy flow of water. Banded tubes, as well as tubes with pitted ornamentation on their walls, were lignified and, when they form single celled conduits, are considered to be tracheids. These, the "next generation" of transport cell design, have a more rigid structure than hydroids, allowing them to cope with higher levels of water pressure. Tracheids may have a single evolutionary origin, possibly within the hornworts, uniting all tracheophytes (but they may have evolved more than once).

Water transport requires regulation, and dynamic control is provided by stomata. By adjusting the amount of gas exchange, they can restrict the amount of water lost through transpiration. This is an important role where water supply is not constant, and indeed stomata appear to have evolved before tracheids, being present in the non-vascular hornworts.

An endodermis probably evolved during the Silu-Devonian, but the first fossil evidence for such a structure is Carboniferous. This structure in the roots covers the water transport tissue and regulates ion exchange (and prevents unwanted pathogens etc. from entering the water transport system). The endodermis can also provide an upwards pressure, forcing water out of the roots when transpiration is not enough of a driver.

Once plants had evolved this level of controlled water transport, they were truly homoiohydric, able to extract water from their environment through root-like organs rather than relying on a film of surface moisture, enabling them to grow to much greater size. As a result of their independence from their surroundings, they lost their ability to survive desiccation – a costly trait to retain.

During the Devonian, maximum xylem diameter increased with time, with the minimum diameter remaining pretty constant. By the middle Devonian, the tracheid diameter of some plant lineages (Zosterophyllophytes) had plateaued. Wider tracheids allow water to be transported faster, but the overall transport rate depends also on the overall cross-sectional area of the xylem bundle itself. The increase in vascular bundle thickness further seems to correlate with the width of plant axes, and plant height; it is also closely related to the appearance of leaves and increased stomatal density, both of which would increase the demand for water.

While wider tracheids with robust walls make it possible to achieve higher water transport pressures, this increases the problem of cavitation. Cavitation occurs when a bubble of air forms within a vessel, breaking the bonds between chains of water molecules and preventing them from pulling more water up with their cohesive tension. A tracheid, once cavitated, cannot have its embolism removed and return to service (except in a few advanced angiosperms which have developed a mechanism of doing so). Therefore, it is well worth plants' while to avoid cavitation occurring. For this reason, pits in tracheid walls have very small diameters, to prevent air entering and allowing bubbles to nucleate. Freeze-thaw cycles are a major cause of cavitation. Damage to a tracheid's wall almost inevitably leads to air leaking in and cavitation, hence the importance of many tracheids working in parallel.

Cavitation is hard to avoid, but once it has occurred plants have a range of mechanisms to contain the damage. Small pits link adjacent conduits to allow fluid to flow between them, but not air – although ironically these pits, which prevent the spread of embolisms, are also a major cause of them. These pitted surfaces further reduce the flow of water through the xylem by as much as 30%. Conifers, by the Jurassic, developed an ingenious improvement, using valve-like structures to isolate cavitated elements. These torus-margo structures have a blob floating in the middle of a donut; when one side depressurizes the blob is sucked into the torus and blocks further flow. Other plants simply accept cavitation; for instance, oaks grow a ring of wide vessels at the start of each spring, none of which survive the winter frosts. Maples use root pressure each spring to force sap upwards from the roots, squeezing out any air bubbles.

Growing to height also employed another trait of tracheids – the support offered by their lignified walls. Defunct tracheids were retained to form a strong, woody stem, produced in most instances by a secondary xylem. However, in early plants, tracheids were too mechanically vulnerable, and retained a central position, with a layer of tough sclerenchyma on the outer rim of the stems. Even when tracheids do take a structural role, they are supported by sclerenchymatic tissue.

Tracheids end with walls, which impose a great deal of resistance on flow; vessel members have perforated end walls, and are arranged in series to operate as if they were one continuous vessel. The function of end walls, which were the default state in the Devonian, was probably to avoid embolisms. An embolism is where an air bubble is created in a tracheid. This may happen as a result of freezing, or by gases dissolving out of solution. Once an embolism is formed, it usually cannot be removed; the affected cell cannot pull water up, and is rendered useless.

End walls excluded, the tracheids of prevascular plants were able to operate under the same hydraulic conductivity as those of the first vascular plant, *Cooksonia*.

The size of tracheids is limited as they comprise a single cell; this limits their length, which in turn limits their maximum useful diameter to 80 μm. Conductivity grows with the fourth power of diameter, so increased diameter has huge rewards; vessel elements, consisting of a number of cells, joined at their ends, overcame this limit and allowed larger tubes to form, reaching diameters of up to 500 μm, and lengths of up to 10 m.

Vessels first evolved during the dry, low CO_2 periods of the late Permian, in the horsetails, ferns and Selaginellales independently, and later appeared in the mid Cretaceous in angiosperms and gnetophytes. Vessels allow the same cross-sectional area of wood to transport around a hundred times more water than tracheids! This allowed plants to fill more of their stems with structural fibers, and also opened a new niche to vines, which could transport water without being as thick as the tree they grew on. Despite these advantages, tracheid-based wood is a lot lighter, thus cheaper to make, as vessels need to be much more reinforced to avoid cavitation.

Development

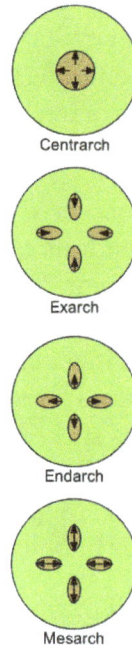

Patterns of xylem development: xylem in brown;
arrows show direction of development from protoxylem to metaxylem.

Xylem development can be described by four terms: centrarch, exarch, endarch and mesarch. As it develops in young plants, its nature changes from *protoxylem* to *metaxylem* (i.e. from *first xylem* to *after xylem*). The patterns in which protoxylem and metaxylem are arranged is important in the study of plant morphology.

Protoxylem and Metaxylem

As a young vascular plant grows, one or more strands of primary xylem form in its stems and roots. The first xylem to develop is called 'protoxylem'. In appearance protoxylem is usually distinguished by narrower vessels formed of smaller cells. Some of these cells have walls which contain thickenings in the form of rings or helices. Functionally, protoxylem can extend: the cells are able to grow in size and develop while a stem or root is elongating. Later, 'metaxylem' develops in the strands of xylem. Metaxylem vessels and cells are usually larger; the cells have thickenings which are typically either in the form of ladderlike transverse bars (scalariform) or continuous sheets except for holes or pits (pitted). Functionally, metaxylem completes its development after elongation ceases when the cells no longer need to grow in size.

Patterns of Protoxylem and Metaxylem

There are four main patterns to the arrangement of protoxylem and metaxylem in stems and roots.

- *Centrarch* refers to the case in which the primary xylem forms a single cylinder in the center of the stem and develops from the center outwards. The protoxylem is thus found in the central core and the metaxylem in a cylinder around it. This pattern was common in early land plants, such as "rhyniophytes", but is not present in any living plants.

The other three terms are used where there is more than one strand of primary xylem.

- *Exarch* is used when there is more than one strand of primary xylem in a stem or root, and the xylem develops from the outside inwards towards the center, i.e. centripetally. The metaxylem is thus closest to the center of the stem or root and the protoxylem closest to the periphery. The roots of vascular plants are normally considered to have exarch development.

- *Endarch* is used when there is more than one strand of primary xylem in a stem or root, and the xylem develops from the inside outwards towards the periphery, i.e. centrifugally. The protoxylem is thus closest to the center of the stem or root and the metaxylem closest to the periphery. The stems of seed plants typically have endarch development.

- *Mesarch* is used when there is more than one strand of primary xylem in a stem or root, and the xylem develops from the middle of a strand in both directions. The metaxylem is thus on both the peripheral and central sides of the strand with the protoxylem between the metaxylem (possibly surrounded by it). The leaves and stems of many ferns have mesarch development.

Phloem

In vascular plants, phloem is the living tissue that transports the soluble organic compounds made during photosynthesis (known as photosynthate), in particular the sugar sucrose, to parts of the plant where needed. This transport process is called translocation. In trees, the phloem is the innermost layer of the bark, hence the name, derived from the Greek word (*phloios*) meaning "bark".

Structure

Phloem tissue consists of: conducting cells, generally called sieve elements; parenchyma cells, including both specialized companion cells or albuminous cells and unspecialized cells; and supportive cells, such as fibres and sclereids.

Conducting Cells (Sieve Elements)

1. Xylem

2. Phloem

3. Cambium

4. Pith

5. Companion Cells

Sieve elements are the type of cell that are responsible for transporting sugars throughout the plant. At maturity they lack a nucleus and have very few organelles, so they rely on companion cells or albuminous cells for most of their metabolic needs. Sieve tube cells do contain vacuoles and other organelles, such as ribosomes, before they mature, but these generally migrate to the cell wall and dissolve at maturity; this ensures there is little to impede the movement of fluids. One of the few organelles they do contain at maturity is the smooth endoplasmic reticulum, which can be found at the plasma membrane, often nearby the plasmodesmata that connect them to their companion or albuminous cells. All sieve cells have groups of pores at their ends that grow from modified and enlarged plasmodesmata, called sieve areas. The pores are reinforced by platelets of a polysaccharide called callose.

Parenchyma Cells

They are of three types:

Companion Cells

The metabolic functioning of sieve-tube members depends on a close association with the *companion cells*, a specialized form of parenchyma cell. All of the cellular functions of a sieve-tube element are carried out by the (much smaller) companion cell, a typical nucleate plant cell except the companion cell usually has a larger number of ribosomes and mitochondria. The dense cytoplasm of a companion cell is connected to the sieve-tube element by plasmodesmata. The common side-well shared by a sieve tube element and a companion cell has large numbers of plasmodesmata.

There are two types of companion cells.

1. Ordinary companion cells, which have smooth walls and few or no plasmodesmatal connections to cells other than the sieve tube.

2. Transfer cells, which have much-folded walls that are adjacent to non-sieve cells, allowing for larger areas of transfer. They are specialized in scavenging solutes from those in the cell walls that are actively pumped requiring energy.

Albuminous Cells

Albuminous cells have a similar role to companion cells, but are associated with sieve cells only and are hence found only in seedless vascular plants and gymnosperms.

Other Parenchyma Cells

Other parenchyma cells within the phloem are generally undifferentiated and used for food storage.

Supportive Cells

Although its primary function is transport of sugars, phloem may also contain cells that have a mechanical support function. These generally fall into two categories: fibres and sclereids. Both cell types have a secondary cell wall and are therefore dead at maturity. The secondary cell wall increases their rigidity and tensile strength.

Fibres

Bast fibres are the long, narrow supportive cells that provide tension strength without limiting flexibility. They are also found in xylem, and are the main component of many textiles such as paper, linen, and cotton.

Sclereids

Sclereids are irregularly shaped cells that add compression strength but may reduce flexibility to some extent. They also serve as anti-herbivory structures, as their irregular shape and hardness will increase wear on teeth as the herbivores chew. For example, they are responsible for the gritty texture in pears.

Function

Unlike xylem (which is composed primarily of dead cells), the phloem is composed of still-living cells that transport sap. The sap is a water-based solution, but rich in sugars made by photosynthesis. These sugars are transported to non-photosynthetic parts of the plant, such as the roots, or into storage structures, such as tubers or bulbs.

During the plant's growth period, usually during the spring, storage organs such as the roots are sugar sources, and the plant's many growing areas are sugar sinks. The movement in phloem is multidirectional, whereas, in xylem cells, it is unidirectional (upward).

After the growth period, when the meristems are dormant, the leaves are sources, and storage organs are sinks. Developing seed-bearing organs (such as fruit) are always sinks. Because of this multi-directional flow, coupled with the fact that sap cannot move with ease between adjacent sieve-tubes, it is not unusual for sap in adjacent sieve-tubes to be flowing in opposite directions.

While movement of water and minerals through the xylem is driven by negative pressures (tension) most of the time, movement through the phloem is driven by positive hydrostatic pressures. This process is termed *translocation*, and is accomplished by a process called phloem loading and *unloading*.

Phloem sap is also thought to play a role in sending informational signals throughout vascular plants. "Loading and unloading patterns are largely determined by the conductivity and number of plasmodesmata and the position-dependent function of solute-specific, plasma membrane transport proteins. Recent evidence indicates that mobile proteins and RNA are part of the plant's long-distance communication signaling system. Evidence also exists for the directed transport and sorting of macromolecules as they pass through plasmodesmata."

Organic molecules such as sugars, amino acids, certain hormones, and even messenger RNAs are transported in the phloem through sieve tube elements.

Girdling

Because phloem tubes are located outside the xylem in most plants, a tree or other plant can be killed by stripping away the bark in a ring on the trunk or stem. With the phloem destroyed, nutrients cannot reach the roots, and the tree/plant will die. Trees located in areas with animals such as beavers are vulnerable since beavers chew off the bark at a fairly precise height. This process is known as girdling, and can be used for agricultural purposes. For example, enormous fruits and vegetables seen at fairs and carnivals are produced via girdling. A farmer would place a girdle at the base of a large branch, and remove all but one fruit/vegetable from that branch. Thus, all the sugars manufactured by leaves on that branch have no sinks to go to but the one fruit/vegetable, which thus expands to many times normal size.

Origin

When the plant is an embryo, vascular tissue emerges from procambium tissue, which is at the center of the embryo. Protophloem itself appears in the mid-vein extending into the cotyledonary node, which constitutes the first appearance of a leaf in angiosperms, where it forms continuous strands. The hormone auxin, transported by the protein PIN1 is responsible for the growth of those protophloem strands, signaling the final identity of those tissues. SHORTROOT(SHR), and microRNA165/166 also participate in that process, while Callose Synthase 3(CALS3), inhibits the locations where SHORTROOT(SHR), and microRNA165 can go.

In the embryo, root phloem develops independently in the upper hypocotyl, which lies between the embryonic root, and the cotyledon.

In an adult, the phloem originates, and grows outwards from, meristematic cells in the vascular cambium. Phloem is produced in phases. *Primary* phloem is laid down by the apical meristem and develops from the procambium. *Secondary* phloem is laid down by the vascular cambium to the inside of the established layer(s) of phloem.

In some eudicot families (Apocynaceae, Convolvulaceae, Cucurbitaceae, Solanaceae, Myrtaceae, Asteraceae, Thymelaeaceae), phloem also develops on the inner side of the vascular cambium; in this case, a distinction between external phloem and internal phloem or intraxylary phloem is made. Internal phloem is mostly primary, and begins differentiation later than the external phloem and protoxylem, though it is not without exceptions. In some other families (Amaranthaceae, Nyctaginaceae, Salvadoraceae), the cambium also periodically forms inward strands or layers of phloem, embedded in the xylem: Such phloem strands are called included phloem or interxylary phloem.

Nutritional Use

Phloem of pine trees has been used in Finland as a substitute food in times of famine and even in good years in the northeast. Supplies of phloem from previous years helped stave off starvation in the great famine of the 1860s. Phloem is dried and milled to flour (*pettu* in Finnish) and mixed with rye to form a hard dark bread, bark bread. The least appreciated was *silkko*, a bread made only from buttermilk and *pettu* without any real rye or cereal flour. Recently, *pettu* has again become available as a curiosity, and some have made claims of health benefits. However, its food energy content is low relative to rye or other cereals.

Phloem from silver birch has been also used to make flour in the past.

Ground Tissue

The ground tissue of plants includes all tissues that are neither dermal nor vascular. It can be divided into three classes based on the nature of the cell walls. Parenchyma cells have thin primary walls and usually remain alive after they become mature. Parenchyma forms the "filler" tissue in the soft parts of plants. Collenchyma cells have thin primary walls with some areas of secondary thickening. Collenchyma provides extra structural support, particularly in regions of new growth. Sclerenchyma cells have thick lignified secondary walls and often die when mature. Sclerenchyma provides the main structural support to a plant.

Parenchyma

Parenchyma is a versatile ground tissue that generally constitutes the "filler" tissue in soft parts of plants. It forms, among other things, the cortex and pith of stems, the cortex of roots, the mesophyll of leaves, the pulp of fruits, and the endosperm of seeds. Parenchyma cells are living cells and may remain meristematic at maturity—meaning that they are capable of cell division if stimulated. They have thin but flexible cellulose cell walls, and are generally polyhedral when close-packed, but can be roughly spherical when isolated from their neighbours. They have large central vacuoles, which allow the cells to store and regulate ions, waste products, and water. Tissue specialised for food storage is commonly formed of parenchyma cells.

Parenchyma cells have a variety of functions:

- In leaves, they form the mesophyll and are responsible for photosynthesis and the exchange of gases, parenchyma cells in the mesophyll of leaves are specialised parenchyma cells called chlorenchyma cells (parenchyma cells with chloroplasts).

- Storage of starch, protein, fats, oils and water in roots, tubers (e.g. potatoes), seed endosperm (e.g. cereals) and cotyledons (e.g. pulses and peanuts)

- Secretion (e.g. the parenchyma cells lining the inside of resin ducts)

- Wound repair and the potential for renewed meristematic activity

- Other specialised functions such as aeration (aerenchyma) provides buoyancy and helps aquatic plants in floating.

- Chlorenchyma cells carry out photosynthesis and manufacture food.

The shape of parenchyma cells varies with their function. In the spongy mesophyll of a leaf, parenchyma cells range from near-spherical and loosely arranged with large intercellular spaces, to branched or stellate, mutually interconnected with their neighbours at the ends of their arms to form a three-dimensional network, like in the red kidney bean *Phaseolus vulgaris* and other mesophytes. These cells, along with the epidermal guard cells of the stoma, form a system of air spaces and chambers that regulate the exchange of gases. In some works the cells of the leaf epidermis are regarded as specialised parenchymal cells, but the modern preference has long been to classify the epidermis as plant dermal tissue, and parenchyma as ground tissue. shapes of parenchyma 1= polyhedral[these cells are most poly hedral shape] 2=stellate (found in stem of plants and have well developed air spaces btween them) 3=elongated(are found in pallisade tissue of leaf) 4=lobed (are found inspongy and pallisade mesophyyll tissue of some plants)

Collenchyma

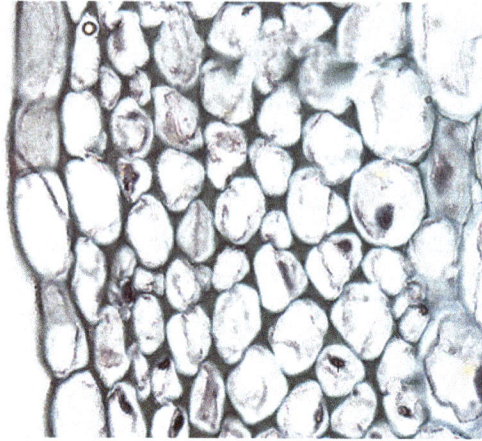

Cross section of collenchyma cells

The first use of "collenchyma" was by Link (1837) who used it to describe the sticky substance on *Bletia* (Orchidaceae) pollen. Complaining about Link's excessive nomenclature, Schleiden (1839) stated mockingly that the term "collenchyma" could have more easily been used to describe elongated sub-epidermal cells with unevenly thickened cell walls.

Collenchyma tissue is composed of elongated cells with irregularly thickened walls. They provide structural support, particularly in growing shoots and leaves. Collenchyma tissue makes up things such as the resilient strands in stalks of celery. Collenchyma cells are usually living, and have only a thick primary cell wall made up of cellulose and pectin. Cell wall thickness is strongly affected by mechanical stress upon the plant. The walls of collenchyma in shaken plants (to mimic the effects of wind etc.), may be 40–100% thicker than those not shaken.

There are four main types of collenchyma:

- Angular collenchyma (thickened at intercellular contact points)

- Tangential collenchyma (cells arranged into ordered rows and thickened at the tangential face of the cell wall)

- Annular collenchyma (uniformly thickened cell walls)

- Lacunar collenchyma (collenchyma with intercellular spaces)

Collenchyma cells are most often found adjacent to outer growing tissues such as the vascular cambium and are known for increasing structural support and integrity.

Sclerenchyma

Sclerenchyma is the supporting tissue in plants. Two types of sclerenchyma cells exist: fibres and sclereids. Their cell walls consist of cellulose, hemicellulose and lignin. Sclerenchyma cells are the principal supporting cells in plant tissues that have ceased elongation. Sclerenchyma fibres are of great economic importance, since they constitute the source material for many fabrics (e.g. flax, hemp, jute, and ramie).

Unlike the collenchyma, mature sclerenchyma is composed of dead cells with extremely thick cell walls (secondary walls) that make up to 90% of the whole cell volume. The term *sclerenchyma* is derived from the Greek (*sklērós*), meaning "hard." It is the hard, thick walls that make sclerenchyma cells important strengthening and supporting elements in plant parts that have ceased elongation. The difference between fibres and sclereids is not always clear: transitions do exist, sometimes even within the same plant.

Fibers

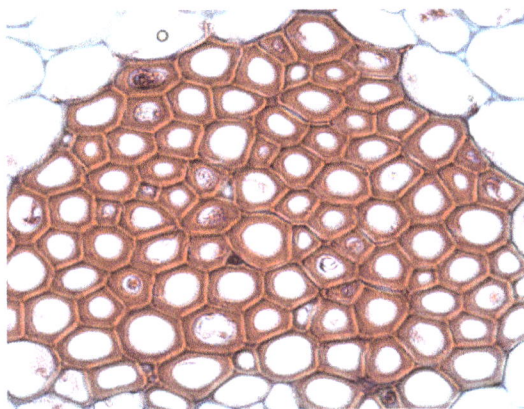

Cross section of sclere nchyma fibres

Fibers or bast are generally long, slender, so-called prosenchymatous cells, usually occurring in strands or bundles. Such bundles or the totality of a stem's bundles are colloquially called fibres. Their high load-bearing capacity and the ease with which they can be processed has since antiquity made them the source material for a number of things, like ropes, fabrics and mattresses. The fibres of flax (*Linum usitatissimum*) have been known in Europe and Egypt for more than 3,000 years, those of hemp (*Cannabis sativa*) in China for just as long. These fibres, and those of jute (*Corchorus capsularis*) and ramie (*Boehmeria nivea*, a nettle), are extremely soft and elastic and are especially well suited for the processing to textiles. Their principal cell wall material is cellulose.

Contrasting are hard fibres that are mostly found in monocots. Typical examples are the fibres of many grasses, agaves (sisal: *Agave sisalana*), lilies (*Yucca* or *Phormium tenax*), *Musa textilis* and others. Their cell walls contain, besides cellulose, a high proportion of lignin. The load-bearing capacity of *Phormium tenax* is as high as 20–25 kg/mm², the same as that of good steel wire (25 kg/mm²), but the fibre tears as soon as too great a strain is placed upon it, while the wire distorts and does not tear before a strain of 80 kg/mm². The thickening of a cell wall has been studied in *Linum*. Starting at the centre of the fibre, the thickening layers of the secondary wall are deposited one after the other. Growth at both tips of the cell leads to simultaneous elongation. During development the layers of secondary material seem like tubes, of which the outer one is always longer and older than the next. After completion of growth, the missing parts are supplemented, so that the wall is evenly thickened up to the tips of the fibres.

Fibres usually originate from meristematic tissues. Cambium and procambium are their main centres of production. They are usually associated with the xylem and phloem of the vascular bundles. The fibres of the xylem are always lignified, while those of the phloem are cellulosic. Reliable evidence for the fibre cells' evolutionary origin from tracheids exists. During evolution the strength of the tracheid cell walls was enhanced, the ability to conduct water was lost and the size of the pits was reduced. Fibres that do not belong to the xylem are bast (outside the ring of cambium) and such fibres that are arranged in characteristic patterns at different sites of the shoot.

Sclereids

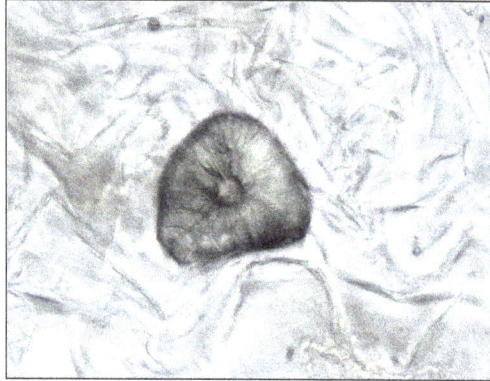

Fresh mount of a sclereid

Sclereids are a reduced form of sclerenchyma cells with highly thickened, lignified walls. These have a shape of a star.

They are small bundles of sclerenchyma tissue in plants that form durable layers, such as the cores of apples and the gritty texture of pears (*Pyrus communis*). Sclereids are variable in shape. The cells can be isodiametric, prosenchymatic, forked or elaborately branched. They can be grouped into bundles, can form complete tubes located at the periphery or can occur as single cells or small groups of cells within parenchyma tissues. But compared with most fibres, sclereids are relatively short. Characteristic examples are brachysclereids or the stone cells (called stone cells because of their hardness) of pears and quinces (*Cydonia oblonga*) and those of the shoot of the wax plant (*Hoya carnosa*). The cell walls fill nearly all the cell's volume. A layering of the walls and the existence of branched pits is clearly visible. Branched pits such as these are called ramiform pits. The shell of many seeds like those of nuts as well as the stones of drupes like cherries and plums are made up from sclereids.

These structures are used to protect other cells.

Long, tapered sclereids supporting a leaf edge in *Dionysia kossinskyi*

Spongy Tissue

Spongy tissue is a type of tissue found both in plants and animals.

In plants, it is part of the mesophyll, where it forms a layer next to the palisade cells in the leaf. The spongy mesophyll's function is to allow for the interchange of gases (CO_2) that are needed for photosynthesis.The spongy mesophyll cells are less likely to go through photosynthesis than those in the palisade mesophyll.It is also the name of a disorder of fruit ripening which can reduce the value of a fruit yield, especially in mango. In the alphonso mango variety, this problem is particularly common, giving soft, white, 'corky' tissue.

References

- Raven, Peter H., Evert, Ray F., & Eichhorn, Susan E. (1986). Biology of Plants (4th ed.). New York: Worth Publishers. ISBN 0-87901-315-X

- Tyree, M.T. (1997). "The Cohesion-Tension theory of sap ascent: current controversies". Journal of Experimental Botany. 48 (10): 1753–1765. doi:10.1093/jxb/48.10.1753

- Koch, George W.; Sillett, Stephen C.; Jennings, Gregory M.; Davis, Stephen D. (2004). "The limits to tree height". Nature. 428 (6985): 851–854. PMID 15103376. doi:10.1038/nature02417

- Editors: Anthony Yeo, Tim Flowers (2007). Plant solute transport. Oxford UK: Blackwell Publishing. p. 221. ISBN 978-1-4051-3995-3

- Tyree, Melvin T. (1997). "The cohesion-tension theory of sap ascent: current controversies". Journal of Experimental Botany. 48 (10): 1753. doi:10.1093/jxb/48.10.1753

- Cronquist, A. (August 1988). The Evolution and Classification of Flowering Plants. New York, New York: New York Botanical Garden Press. ISBN 978-0-89327-332-3

- McCulloh, Katherine A.; John S. Sperry; Frederick R. Adler (2003). "Water transport in plants obeys Murray's law". Nature. 421 (6926): 939–942. Bibcode:2003Natur.421..939M. PMID 12607000. doi:10.1038/nature01444

- Taylor, T.N.; Taylor, E.L.; Krings, M. (2009). Paleobotany, the Biology and Evolution of Fossil Plants (2nd ed.). Amsterdam; Boston: Academic Press. pp. 207ff., 212ff. ISBN 978-0-12-373972-8

- Lucas, William, et al. The Plant Vascular System: Evolution, Development and Functions. Journal of Integrative Plant Biology. 55, 294-388 (2013) PMID 23462277

- Turgeon, Robert; Wolf, Shmuel (2009). "Phloem Transport: Cellular Pathways and Molecular Trafficking". Annual Review of Plant Biology. 60: 207–21. PMID 19025382. doi:10.1146/annurev.arplant.043008.092045

- Evert, Ray F; Eichhorn, Susan E. Esau's Plant Anatomy: Meristems, Cells, and Tissues of the Plant Body: Their Structure, Function, and Development. Publisher: Wiley-Liss 2006. ISBN 978-0-471-73843-5

- Niklas, K.; Pratt, L. (1980). "Evidence for lignin-like constituents in Early Silurian (Llandoverian) plant fossils". Science. 209 (4454): 396–397. Bibcode:1980Sci...209..396N. PMID 17747811. doi:10.1126/science.209.4454.396

Permissions

Index